1996

Nuclear Energy

Nuclear Energy

Simon Rippon

Foreword by
Sir John Hill

Tameside Public Libraries

769620

HEINEMANN: LONDON

Willian Heinemann Ltd
10 Upper Grosvenor Street London W1X 9PA
LONDON MELBOURNE TORONTO
JOHANNESBURG AUCKLAND

First published 1984

© Simon Rippon 1984

434 91731 1

Printed in Great Britain by
The Pitman Press, Bath

Contents

	Foreword	vii
	Preface	ix
1	The Questions They Ask	1
2	Nuclear Reactors	21
3	Nuclear Fuel	44
4	Nuclear Safety	68
5	Politics of Nuclear Energy	91
6	Waste Management	105
7	Advanced Reactors	128
8	Nuclear Industry Worldwide	152
	Glossary	199
	Index	210

Foreword

The principal difference between the nuclear industry and other industries with similar technological and engineering characteristics lies in the field of public interest in and public perception of its workings. Members of the public are not really well-informed about the nuclear industry, neither would it be reasonable to expect them to be so. They are similarly not particularly well-informed about the workings and problems of other energy-generating industries and rightly do not regard such information as necessary. They are in general concerned about how industries affect them but not how the factories are designed and operated.

There are however some features about nuclear power that have caught public attention in a way that stands it apart from other industries. The ability to produce massive supplies of energy from a small piece of uranium has a fascination for all. The ability of a submarine to sail round the world several times without refuelling is a measure of power in reserve. The fact that radiation could do you harm without your being aware of it or able to feel it is worrying to many. The fact that nuclear power was born out of a weapon of great magnitude raises the question of whether similar consequences could happen in a nuclear plant by accident.

For these and perhaps other reasons we saw in the middle and late 1970s in many of the developed countries of the Western world a polarization of those who had become opposed to power against the nuclear industry and those who supported them. The debate and the arguments deployed has not been particularly rewarding from the point of view of the public becoming better informed about the issues involved. Public opinion research in the United States has shown that substantial sections of the public there overestimate the hazards of nuclear power by huge factors – over 10 000 times – by comparison with the best statistical analyses that can be made.

Nuclear Energy

On the other hand the nuclear industry is not an easy one to build or to operate to the very high standards required. In a real world with real people who can be fallible it is not enough to reply that everything has been checked and double-checked and nothing can go wrong. But neither does the public get much comfort from the purely academic scientific approach that concludes that the probability of nuclear power destroying the solar system is very small.

The unjustified and emotional attacks that have been made on the nuclear industry have made it more difficult to discuss rationally and objectively many of the manageable but nevertheless real difficulties that the industry has to tackle.

My own view is that the nuclear industry is a safe industry that should give the public less cause for concern than many other industries that are cheerfully accepted by our society. It has a good record and can be proud of much, but not everything, it has done. It is composed of well-educated, dedicated and responsible people, but that does not mean that all decisions have been right or that there have not been lapses at times.

Some of the presentations to the public of the risks of accidents or contamination of the environment have been exaggerated to the point of irresponsibility, but some of the materials being handled are very unpleasant, presenting the industry with many problems that will take time and money to solve.

This book gives a comprehensive review of the development and evolution of nuclear power world wide. It surveys the types of nuclear power station and nuclear fuel plants that have been developed and built over the years. It discusses the underlying science and technology. It discusses the problems that the industry has to face – scientific, engineering, environmental and political. It discusses perhaps for the first time the importance of factors such as industrial structure, market size and the requirement for replication if the industry is to operate efficiently and economically. Finally, it outlines the technological status of the fast reactor and fusion which, when the need arises, can assure the world of continued electricity supplies for as far ahead as it is profitable to speculate.

Sir John Hill

Preface

As a schoolboy in the late 1940s I decided that I ought to find out something about atomic energy and read a fairly basic book about the atom and its energy. I was fascinated in the first place by the way in which most of the physical and chemical phenomena that I had been learning about in science classes could be related to the behaviour of the atoms that make up all matter – today, of course, most science teaching starts with a basic understanding of atomic physics. But I was also intrigued by the concept of vast amounts of energy locked up in the nucleus of the atom and excited by the prospect of this nuclear energy being released in a useful form for the benefit of mankind rather than his destruction.

I mention this personal experience because I think that most of the scientists and engineers who have been responsible for progress from the early scientific beginnings to a major energy supply industry were probably motivated by a similar desire to contribute to the development of nuclear power for the good of mankind.

More recently we have seen the emergence around the world of environmentalists, no doubt equally motivated by a desire to contribute to a better world, but they have chosen – mistakenly, I believe – to devote a large amount of their effort to attacking nuclear energy. This conflict on a subject of considerable complexity has confused and worried many members of the public.

Thus it is that a book on nuclear energy can no longer confine itself to describing the fascinating basics of the technology but must address the broader social implications. It is no bad thing that technologists should submit to such examination, but there is also a need for a better informed public to influence future decisions if the world is to avoid stagnation and eventual decline.

This book aims to give an initial overview of the issues surrounding the civilian development of nuclear energy and to follow this up with further explanation of the main topics. It is not intended to convert the reader into a nuclear engineer, but in understanding better what it is all about he may be able to form a better judgement of which experts to believe. Hopefully too, some student readers may find sufficient interest in the technology to want to pursue their studies further in this direction and contribute to realization of the considerable potential that still exists to obtain wider benefits from nuclear energy.

Simon Rippon

1

The Questions They Ask

What sort of world do we want to live in? This question is frequently posed by the proponents of a so-called 'post-industrial society' in which we all live a simpler life consuming less energy. But technologists and industrialists, contrary to the image that has been created in recent years, are also motivated by their perception of a better world for all of us. It is true that most books on a particular industry relegate social and environmental questions to the end but this is stimulated more by enthusiasm to describe the technology than lack of awareness of the basic social questions. The technologist is as aware as anybody of the shortcomings of present industrialized society and shares the disappointment of the public when new technologies do not live up to initial expectation. Industrial societies have, however, generated a vast improvement in the general standard of living over the past hundred years and technologists would argue that the potential still exists for further real improvement in the quality of life through better application of advanced technology rather than turning back to a post-industrial society.

Although individual views of the sort of world in which we want to live will vary in detail and may change with changing circumstances, it is reasonable to assume that high on the list of priorities are world peace, national stability, and personal security. An adequate supply of energy is an important factor, though certainly not the only factor, in achieving all three of these objectives.

Politicians may justify past wars on ideological grounds but examination of history reveals that all too often the final military showdown has been sparked off by far more down-to-earth considerations of material self-interest. It is difficult in today's world to see anything that is more likely to stimulate such actions than the need to secure energy supplies. It may be that in the longer term food and other vital commodities will assume comparable importance. If, however, we can overcome the problem of energy supplies we will be better placed to deal with the longer-term problems, not least because energy is an important factor in advanced methods for the recycling of raw materials and more efficient production of food.

At the national level economic wellbeing is a major factor in political stability, and here again it is difficult to see anything in the next few decades that is going to be more critical than the availability of an adequate supply of energy at a price which will allow countries to maintain a competitive position in world markets. And nations with strong economies are better able and usually more willing to provide aid to the developing world.

There is still a huge disparity between the haves and the have-nots, both globally and at the individual national level. Most of us would like to see the gap being closed; many would support the view that the haves could manage with a little less, but few would accept that they personally were the ones to manage with less. While a small number of idealists, speaking usually from a position of relative affluence, may advocate a simpler lifestyle, the vast majority of people are very wary of any erosion in their standard of living and a great many would still like to see a considerable improvement.

The concern of individuals is stimulated by the apparent trend towards less material wellbeing at greater cost. In seeking technologies that will help us to halt or even reverse this trend we want therefore to concentrate on greater efficiency – more from less. This is a basic characteristic of nuclear energy. It yields much more energy from less natural resources at less cost. Subsequent chapters of this book will attempt to establish these facts both in terms of financial costs and environmental and social costs.

How much energy do we need?

How much energy do we need in industrialized countries to maintain standards of living and improve them for the less privileged members of society? How much does the Third World need to improve its lot and close the gap in standard of living with the industrialized world? And perhaps the most important question: how short of energy do we have to be before we start fighting over it?

It is impossible to project future requirements precisely and forecasting in the past has been notoriously inaccurate. But it is important to keep on trying to make the best estimate of future energy needs because the provision of new sources of energy requires long-term planning.

The Questions They Ask

Sceptics say, with some historic justification, that the only thing you can be sure of with energy forecasts is that they will turn out to be wrong. Individual national forecasts made in the 1960s and early 1970s on the basis of historic trends and a classical relationship between gross domestic product and energy consumption, mostly assumed an over-optimistic projection of economic growth. Since the forecasts of international organizations have, for the most part, been based on compilations of data from national bodies they too have tended to overestimate global energy requirements.

In recent years, however, the much greater awareness of energy problems has led to more sophisticated forecasting, both at national and international levels. Even so these are not producing greater precision in future projections but are rather generating a wide range of possible scenarios which highlight the very great uncertainty about real energy needs. There now appears to be an inclination among planners to assume that because we overestimated in the past it is now wise to choose the lowest of the range of future energy projections in making decisions. This is rather disturbing. Overestimating of energy needs may have caused investment in power plants ahead of requirements – which is not so disastrous, especially in a period of rapid inflation – but underestimating of future requirements could result in severe shortages.

Having said this, the trends in future energy demand shown in Figure 1.1 are based on a low growth projection, in order to illustrate that

Figure 1.1 Past and likely future pattern of world energy consumption

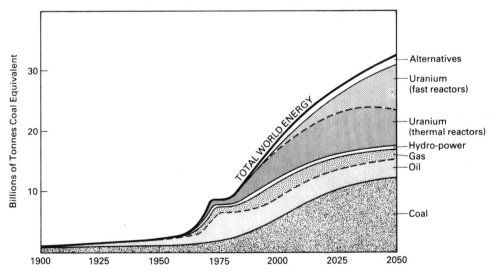

globally this will still involve a very considerable increase in total needs. An average rate of economic growth of around 2 per cent per year has been assumed. The developing world will in fact be hoping for a considerably faster rate of growth, which implies a lower rate in the industrialized world – something that we may have to accept but not a future that any political leaders would dare to contemplate if they are to win popular votes in an election.

It is interesting to note how low our energy consumption was at the beginning of the century. Most of the requirements were met with coal. In the first fifty years the growth in demand was modest and most of the increase was supplied by new discoveries of oil. After the Second World War, however, there was a dramatic increase in energy consumption, largely supported by relatively cheap oil. It has been accompanied by a very significant increase in the standard of living, especially in Western Europe and Japan, and by a large growth in private and commercial motor transport.

The oil crisis of 1973 and the sharp rise in oil prices that followed, halted the runaway growth of energy demand, partly due to the world economic recession, and also due to efforts to use energy more efficiently. In the future it is hoped that greater conservation of energy will change the character of the demand curve from the increasing upward trend in the 1950s and 1960s to a flattening of growth in demand. But it is still important to recognize that there is a very substantial underlying growth in real energy demand, due to the anticipated increase in world population, the aspirations for improvements of living standards in the developing world and the hope of the industrialized world that there may be some recovery in the general economic climate. The approximate doubling in demand between now and the end of the century, followed by a further doubling in the first half of the next century, as in Figure 1.1, is not excessive when one takes account of increases in world population. This is expected to double before 2025 and may level off at two-and-a-half times the present level after 2050.

Do we need nuclear energy?

The pattern of energy consumption suggested in Figure 1.1 indicates a 25 per cent increase in the consumption of oil and gas by the end of the century. It may be optimistic to assume that such increases will be available and the decline of supplies beyond that seems inevitable. It is also postulated that coal production will be increased by about 50 per cent by the end of the century and after that there could be a very large increase in demand for coal to substitute for oil and gas. This assumes that by the turn of the century technology will have been developed for large-scale economic conversion of coal to gas and synthetic liquid fuels. There is little potential for increasing the contribution of hydroelectric power. A variety of alternative sources of energy such as solar, wind, tidal and wave, have been attracting a lot of attention and may by the turn of the century be able to make a small contribution of around 5 per cent to total energy requirements, but the intermittent nature of these energy sources means that they can only be considered as supplementary supplies. There is a long-term possibility of virtually unlimited supplies of energy some time in the next century from nuclear fusion but the state of development of this very advanced technology (which will be described in Chapter 7) is still too uncertain to enable us to predict when it may start to make a significant contribution.

The only other source of energy which is available today with the potential for appreciably expanded use is nuclear energy. A three- or four-fold increase in nuclear power plant capacity by the end of the century might just fill the gap between supply and demand, although it should be remembered that the projections of demand are very conservative and assume a vigorous parallel effort in the conservation of energy.

At first sight the contribution of nuclear energy to the total primary energy consumption seems to be relatively small – around 7 per cent in 1980 and 18 to 20 per cent in 2000. Nuclear energy is, however, used in the vital sector of electricity production. A country's level of industrial activity is increasingly dependent on reliable and economic supplies of electricity and many essential services also depend on electricity. By replacing oil in the prime area of electricity production nuclear energy is already making a very valuable contribution and its expanded use could contribute to stabilization of oil prices. In addition, the increased use of nuclear energy for electricity production in industrialized countries will release some oil onto the world market and could help the developing world where advanced nuclear technology is at present less appropriate.

A perspective on the valuable contribution that nuclear energy is already making in substituting for oil can be obtained from a couple of examples. Nuclear power plants currently operating in the United States produce between 11 and 12 per cent of electricity needs, which is equivalent to about 90 million tonnes of oil per year. This is roughly equal to the total annual oil consumption in Britain. Likewise nuclear plants in Britain currently produce between 12 and 14 per cent of electricity needs, which is equivalent to 7 million tonnes of oil per year, which is more than enough to meet the oil requirements of many developing countries.

In the next ten to twenty years the main role of nuclear energy will be to replace oil in electricity generation and so help to reduce the dependence of the United States, Western Europe and Japan on imported oil. In the longer term, as coal is needed to substitute for oil and gas in other vital areas, nuclear energy is likely to start taking over the electricity production currently met by coal-fired power stations. Nuclear energy also has the potential to supply heat directly and, if it is allowed to, could play a useful role in the so-called 'process heat' sector. Applications could include district heating, gasification of coal, production of drinking water from the sea, and the supply of hot gases for steel-making.

Do we have enough uranium?

It can be seen from Table 1.1 that nuclear energy certainly produces far more electricity from less natural resources than any other established source of energy. It is also noted that in the present types of commercial nuclear power plants, which use the so-called 'thermal reactors', only about 1 per cent of the uranium fuel is actually used up, and clearly the possibility exists to get even more energy from yet smaller amounts of uranium in more efficient types of plants. This fact was recognized from the earliest days of civil nuclear power plant development and a great deal of work has been done in several different countries to develop a more advanced system known as the 'fast reactor'. While producing power in its central core, the fast reactor also has the capability to convert unused uranium in a blanket around the core into an efficient new fuel material – plutonium – which can be recycled in the core to produce more power. This process is referred to as *breeding*. Fuel recycling and

Table 1.1

Quantity of different fuels needed to generate one billion* kilowatt-hours (or units) of electricity	
Coal	124 000 tonnes
Oil	83 000 tonnes or 553 000 barrels
Natural gas	96 million cubic metres
Hydropower	3.7 billion tonnes of water falling 100 metres
Uranium – with 1% utilization in present thermal reactors	4.9 tonnes
Uranium – with 75% utilization in fast reactors	65 kilogrammes

*1 billion = 1000 million

fast reactors are described in greater detail in Chapters 3 and 7. At this stage it is only necessary to appreciate that these techniques allow one to use up about 75 per cent of the original uranium and therefore to obtain much more energy – about fifty to sixty times as much – from a given amount of uranium.

Known reserves of uranium in the world should certainly be sufficient, even if they are used only in the relatively inefficient thermal reactors, to supply the presently envisaged nuclear power plants well into the next century, and there are reasonable prospects that more uranium will be discovered. But two factors could cause the uranium supply situation to become more difficult around the turn of the century. The first of these would be a more rapid expansion of nuclear power programmes in the 1990s in response to further worsening of the oil supply situation or failure to achieve anticipated expansion of coal production. The other factor is political. With the exception of the United States and the Soviet Union, the main uranium consuming countries are different from the main producing countries and there are already some signs that uranium supplies could be influenced by international politics in the future just as oil supplies are today.

The way to avoid possible difficulties with uranium supplies is to make sure that the fast reactor is ready for commercial use in the 1990s. The importance of this for national self-sufficiency has already been recognized by some leading political figures. In France, for example, it has been noted that the country's modest reserves of uranium, which currently supply about half the requirements of the present nuclear power plants, could, if used in fast reactors, represent an energy source equivalent in size to the oil reserves of Saudi Arabia.

Is nuclear energy safe?

Shortly after the discovery of radioactivity at the end of the nineteenth century it was recognized that nuclear radiation offered both benefits in medical diagnostics and treatment and risks from overexposure. As early as 1928 an international commission of experts in the field was established to keep under review all research work into the harmful effects of radiation and to make recommendations on safety levels for those working with radiation. Today we probably have a better understanding of the harmful effects of radiation than of any other environmental pollutant and the recommendations of the International Commission on Radiological Protection form the basis of strict legal limits on levels of radiation exposure in virtually every country of the world.

Nuclear Energy

Therefore, when nuclear energy was introduced for peaceful production of power in the 1950s it was with a full appreciation of the potential hazards of the radioactive materials generated inside nuclear reactors. In this respect the birth of the nuclear industry was different to the historical development of other industrial activities where the associated dangers were frequently not recognized until there had been serious accidents. From the outset, nuclear reactors have been built and operated with two basic safety principles in mind. The first of these is the so-called 'fail-safe' principle in which, if there is a failure of equipment, the plant will automatically shut down safely. The second is the provision of many back-up layers of protection, sometimes referred to as *defence in depth*, to prevent, or greatly reduce, the consequences to plant operators and the public if there is an unpredicted failure or combination of failures.

As a result of this early awareness of safety, the record of the nuclear industry worldwide has to date been exceptional. There have been no deaths directly attributable to radiation after something like 2500 reactor-operating years at civil nuclear power plants. There have been a few isolated cases of argument about former workers in nuclear fuel processing plants who have contracted cancer in later life and, although it is impossible to say definitely whether these cases have been the result of natural causes or radiation, awards have been made on the benefit of the doubt. More important, perhaps, is the fact that the overall safety record of the nuclear industry for all types of accidents is better than the average of other industrial activities and significantly better than many other activities associated with energy supply. Some figures produced from official statistics in Britain, where the nuclear industry is more than twenty-five years old, are shown in Table 1.2.

Later chapters will describe in more detail the safety measures applied at every stage of nuclear energy production and in the different types of

Table 1.2

Estimated number of deaths due to accidents for each 1000 megawatts of electricity generated per year in the United Kingdom

Coal	Extraction	1.4		
	Transport	0.2	Total	1.8
	Generation	0.2		
Oil and gas	Extraction	0.3	Total	0.3
Nuclear	Extraction of uranium	0.1		
	Generation and fuel processing	0.15	Total	0.25

8

plants. It would, of course, be too much to expect that readers will be able to acquire sufficient understanding of all the technical details to make their own judgement of whether or not a particular plant is safe enough. This we have to leave to independent teams of highly qualified experts. Such teams of experts have been established in every country in which nuclear power plants have been, or are being, built. Typically they are likely to take between two and five years, with many hundreds of man-years of effort, to assess the detailed design of a plant before issuing a construction permit. They specify stringent quality control standards to be met in manufacture of all the component parts and undertake a further detailed assessment before issuing operating licenses. And throughout the operating life of plants the licensing authorities still insist on routine inspection and frequent minor improvements in the light of the growing fund of operating experience from around the world.

But what if . . .?

Contrary to the popular image of the arrogant technocrat which seems to have become fairly widespread in recent years, scientists and engineers have never considered themselves to be infallible. They accept that accidents do happen, usually from a combination of equipment failure and human error, and they spend a great deal of effort and money in providing protective systems to minimize the consequences of postulated accidents, even though they may not be able to specify precisely how such accidents could come about. In the case of nuclear power plants this takes the form of many different barriers designed to prevent release of large amounts of radioactivity into the environment. But many people still have the impression that if, in spite of all the protective barriers, a major nuclear accident did occur the consequences would be far more terrible than anything we have experienced, or are likely to experience, from other human activities or natural disasters. Detailed analysis of risks suggests, however, that this is not the case.

The most likely consequence of a serious nuclear accident – one involving very severe damage to the reactor at the heart of a nuclear power plant – is that nobody either working at the plant or among the general public would be injured. This was the case with the accident at the Three Mile Island plant near Harrisburg in the United States in 1979, with the accident which destroyed the interior of a small reactor at Lucens in Switzerland in 1969, and with the severe accident in an early military reactor at Windscale in Britain in 1957.

Before going on to speculate about the worst imaginable consequences it is worth comparing the most likely consequences with those of serious accidents in other areas. The most likely consequence of a serious aircraft accident is that all the passengers and crew will be killed. The most likely consequence of a serious accident at a large chemical complex or oil refinery is that ten to a hundred people working around the plant will be killed or severely injured. If now we consider a serious nuclear accident, which is improbable in the first place, followed by the worst possible combination of circumstances such as escape through the protective barriers of an appreciable proportion of the radioactivity in the reactor, location of the plant in a semi-urban area, and the wind blowing towards the nearest centre of population, then it is possible to postulate several thousand deaths within a few weeks and maybe some tens of thousands of premature deaths from cancer spread over a period of several decades after the accident. But to put these worst imaginable consequences into perspective, they should be compared with the worst imaginable disasters in other areas of human activity. An aircraft crashing on a large stadium could kill tens of thousands. With very unfavourable wind conditions and the release of poisonous gases, a serious accident at a chemical plant could kill hundreds or even thousands of people. And failures of large dams have been the cause of hundreds and thousands of deaths.

But it is of little consolation to establish that the worst imaginable consequences of a nuclear accident are no greater than accidents caused by other activities of man unless we also establish that the probability of such combinations of circumstances is extremely small in all cases. A technique known as risk analysis has been used in a number of countries in an attempt to estimate the probability of various degrees of seriousness of accidents in nuclear plants and the results have been compared with other man-made and natural disasters. Some figures taken from a very extensive risk analysis undertaken in the United States are shown in Table 1.3. The nature of risk analysis is such that these probability figures are very approximate but they do establish the fact that the probability of causing thousands of deaths is extremely small in all the cases considered and much less in the case of nuclear accidents than in other man-made disasters.

What about waste?

Like most other industrial activities, the operation of nuclear power plants gives rise to the generation of waste products and careful control of these wastes is of vital importance. (The various procedures used for the management of different kinds of nuclear waste will be considered in Chapter 6.) It is worth trying to gain a clearer perspective of the

Table 1.3

Approximate probability for different sizes of accidents from various man-made and natural causes compared with the risk from one hundred nuclear power plants operating in the United States

Type of accident	Tens of deaths	Hundreds of deaths	Thousands of deaths
Air crashes	4 per year	1 in 3 years	1 in 3000 years
Fires	1 per year	1 in 6 years	1 in 300 years
Explosions	1 in 3 years	1 in 20 years	1 in 100 years
Dam failures	1 in 10 years	1 in 20 years	1 in 60 years
Nuclear accidents	1 in 2000 years	1 in 10 000 years	1 in 1 million years
Tornados	1 per year	1 in 4 years	—
Hurricanes	1 in 2 years	1 in 6 years	1 in 20 years
Earthquakes	1 in 10 years	1 in 20 years	1 in 40 years
Meteors	1 in 10 000 years	1 in 100 000 years	1 in 1 million years

magnitude of the problem because the preoccupation of critics with this subject has created the false impression that: nuclear power plants produce lots of waste; nuclear waste is much nastier than other kinds of waste; and nuclear waste is dangerous for much longer than other kinds of waste.

A crude estimate of how much waste is produced by any kind of power station can be derived from the amount of fuel it consumes. For all practical purposes the total weight of fuel consumed in the power plant is converted into a roughly equal weight of waste products. Consideration must be given to the safe management of these waste products whether it be by dispersion, as in the case of flue gases and very low levels of radioactivity, or isolation, as in the case of ashes and filtered soot or spent nuclear fuel. Thus the comparison of the amount of fuel required to produce a billion kilowatt-hours of electricity shown in Table 1.1 (page 6) gives a rough guide to the relative amounts of waste produced by different power stations. The situation is complicated slightly by the fact that some radioactive contamination is also induced in materials other than the nuclear fuel in and around a reactor. But associated material is also converted into waste on a much larger scale in plants burning coal, oil or gas because the huge amounts of oxygen used to support combustion are converted to carbon dioxide. The important thing to recognize is that nuclear power plants produce very large amounts of energy from a very small amount of fuel and it is therefore practical to contemplate waste management procedures which will ensure virtually total isolation from the environment without incurring an excessive economic penalty.

Very intense levels of nuclear radiation can cause a type of severe burning and more or less instantaneous death in much the same way as exposure to intense heat radiation from a large fire. In the case of spent fuel which is taken out of a nuclear reactor after one to three years of power production, or the highly concentrated waste materials which can be separated from the spent fuel by chemical reprocessing, the intensity of radiation could still be high enough to cause burning and probable death after ten years. In this respect the waste is considerably nastier than the hot ashes discharged from a coal-fired plant. But if, for the sake of argument, the nuclear wastes were diluted by mixing with a large volume of concrete comparable to the quantities of ashes from the coal plant, the direct radiation would be reduced to fairly harmless levels from the start. In practice it makes more sense to encapsulate the highly concentrated nuclear wastes in very much smaller volumes and surround the capsule with a layer of concrete shielding, because it only takes a metre or two of such protection to reduce the direct radiation to a safe level.

The Questions They Ask

Lower intensities of radiation exposure, or ingestion of small quantities of radioactivity into the body, are unlikely to cause direct fatality but can increase the chances of an early death from cancer. There is also a smaller, but significant, possibility of genetic damage. It is for these reasons that the greatest effort in waste management is devoted to methods to isolate radioactive waste materials and to ensure that they do not find their way back to man even by the most devious route. It should, however, be recognized that there are many substances in the wastes from conventional power stations which are toxic and can cause cancer or genetic damage. These include huge volumes of gases such as sulphur dioxide and oxides of nitrogen, heavy metals such as cadmium, mercury and lead, and even small amounts of naturally occurring radioactive materials. Due to the relative differences in the quantities of wastes for a given amount of power production, the total lethal potential contained in the conventional wastes is at least as great as that from nuclear wastes, but in this case the risk to man is still able to be reduced to an acceptable level by widespread dispersion of the waste products in the environment.

Residual plutonium, which can constitute about 1 per cent of nuclear wastes, has been labelled by the critics of nuclear power as 'the most toxic substance known to man'. The figures in Table 1.4 show that it is certainly very poisonous but the comparisons with other materials indicate that it is not necessarily the worst. It should be mentioned that plutonium

Table 1.4

Comparison of plutonium with some other highly toxic materials

	Lethal dose (milligrammes)	Time to death
INGESTION		
Anthrax spores	under 0.0001	
Botulism	under 0.001	
Lead arsenate	100	hours to days
Potassium cyanide	700	hours to days
Plutonium	1 150	over 15 years
Caffeine	14 000	days
INHALED		
Plutonium	0.26	over 15 years
Nerve gas	1.0	few hours
Plutonium	1.9	1 year
Benzpyrene (in one pack per day of cigarettes for 30 years)	16	over 30 years

is a very efficient nuclear fuel and there is an incentive to separate as much as possible from the waste materials so that it can be recycled and eventually consumed to produce more energy. There will, of course, be small traces of plutonium left in the wastes after the bulk has been recycled but it is a little difficult to conceive how this plutonium, which has defied the best possible chemical separation processes, will get concentrated by natural processes and find its way back to man in dangerous quantities.

The question of how long nuclear waste is dangerous is the one about which there is a great deal of misconception. All radioactive substances decay and, indeed, it is only as the individual atoms decay that they emit radiation. The measure of the rate of decay is known as the half-life and is the time taken for half the atoms in a given mass of material to decay. Substances with a very long half-life, such as naturally occurring uranium with a half-life of billions of years, decay very slowly and therefore emit a relatively low level of radiation from a given number of atoms. Radioactive substances with a short half-life decay quickly and initially emit more radiation from the same number of atoms. But as this number of atoms decays, the level of radiation emission drops off more rapidly and will eventually fall below the radiation emission from the long half-life substance.

In the fission process, from which nuclear energy is produced, some uranium atoms are split into a variety of radioactive atoms with much shorter half-lives than the uranium – in the range from a few seconds to a few hundreds of years. The level of radiation emission from these fission product atoms is initially much greater than that from the uranium atoms which were destroyed in their production but eventually decays below the level that would have been emitted by the uranium. From a very detailed analysis of the radioactive substances likely to be produced by fission of uranium and other materials such as plutonium which are produced from uranium in different side reactions, it is possible to compare the decay of the radioactivity from the waste products with the radioactivity of the uranium ore used as fuel in the first place. Such a comparison is shown in Figure 1.2 and indicates that in the first five hundred years the radiation emission will fall off rapidly to a level close to that of the original uranium and between one thousand and ten thousand years the level will decay below that of the uranium.

Specialists involved in waste management analyse in very great detail the precise hazard of individual constituents of nuclear wastes but the rough comparison with uranium provides a simple indication of how long we need to worry about the problem. Bearing in mind that many generations of people have lived in houses built from rocks containing a

The Questions They Ask

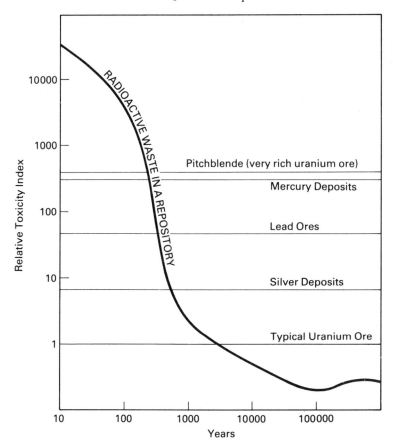

Figure 1.2 Decay of the relative toxicity of radioactive waste in comparison with the toxicity of some typical natural ore bodies

significant amount of uranium or drunk water that has flowed over uranium-containing rocks without apparent harm, it can be judged that if we ensure greater isolation of nuclear waste for up to 1000 years it is certainly not going to represent a major hazard to man. In practice a variety of schemes which have been put forward for the final disposal of highly concentrated wastes offer a degree of isolation which will be a great deal better for much longer.

Essentially what is proposed is a number of different barriers: first the waste will be incorporated in solid blocks of glass which would take several thousand years to leach away even if water were to come into contact with the glass surface; then the blocks of glass could be put into a container of a corrosion-resistant material which could resist water for hundreds or thousands of years; around the containers in a final repository there would be additional layers of an absorbent clay which would mop up any ground water and much of the radioactive material if it did eventually get through the first two barriers; and the repository would be located several hundred metres down in a stable geological formation such as rock salt, granite, clay or deep layers of silt found on ocean beds. Even if some radioactive material were to escape through these barriers there is no reason for it to make a bee-line for the nearest population centre. Rather it would be greatly diluted in diffusion through the ground or in the ocean and at ground level it would be further diluted in the environment.

What about proliferation?

When the Atoms for Peace programme was launched by President Eisenhower back in the mid-1950s, it was widely hailed as the finest example of turning swords into ploughshares in the history of mankind. At that time all the basic technology required to exploit the peaceful uses of nuclear energy was made freely available and substantial aid was provided for the launching of research and development activities in many parts of the world. The only area of technology which remained classified was the enrichment of uranium which could be used for the production of the very high concentration of the fissile uranium isotope used in nuclear explosives. There was little concern at this time with the plutonium which is produced as a by-product in all nuclear reactors and can be separated as an almost pure fissile material by relatively straightforward chemical processes.

A United Nations agency – the International Atomic Energy Agency (I.A.E.A.) based in Vienna – was, however, established with the dual role of promoting the use of nuclear energy for peaceful purposes and establishing a system of safeguards to ensure that there was no diversion of fissile materials for military purposes. Later, in 1968, when the Non Proliferation Treaty (N.P.T.) agreement was concluded, the I.A.E.A. system of international safeguards was adopted as a means of policing compliance with N.P.T. Today, 110 nations have signed the N.P.T. and the I.A.E.A. has concluded safeguards agreements with most of them to allow inspectors to check the activities at most nuclear installations. The system is not yet, and perhaps never can be, watertight but it is already unique in the history of international relations in the degree of control that it allows over national sovereignty and every effort should be made to build on this success rather than highlighting the remaining weaknesses.

The Questions They Ask

There is no doubt that a large number of countries now have the technical knowledge to be able to make a nuclear explosive device, whether it be of the type using highly enriched uranium or plutonium. Whether they have the huge financial resources that would be needed to develop this knowledge into an effective military capability is another question. What is very much more doubtful is whether the parallel development of nuclear power for civil purposes has been, or could be, of any great importance in accelerating the proliferation of nuclear weapons capability. No nuclear weapons have been made anywhere in the world from material produced in civil nuclear power plants and there are good reasons for believing that they never will be because there are cheaper and quicker routes. The existence of civil nuclear power plants in a country can actually be a positive disincentive to nuclear weapons production. This is because the nuclear power plants are likely to become vital sources of cheap electricity for the country concerned but development of nuclear weapons is likely to cause other countries to stop supplying the essential fuel and supporting services for the nuclear power plants.

Such a situation has arisen in India where a nuclear explosive, ostensibly for non-military applications such as excavation or enhanced extraction of oil, gas and other minerals, has been tested. In this case, material for the nuclear explosive was produced in a research reactor, but as a result of the test the Canadians have stopped supplies of heavy water and the Americans are hesitating about further supply of enriched uranium fuel for India's civil nuclear power plants.

In the final analysis the means to prevent nuclear weapons proliferation is to work for a more peaceful world in which nations do not need to rely on such a devastating deterrent. A small but important contribution to achieving a more peaceful world is to ensure that all countries have access to adequate supplies of energy and the civil development of nuclear power can certainly help in this direction. In the meantime considerable international effort is being devoted to the establishment of more effective methods of control to ensure that the peaceful uses of nuclear energy are not abused – some of these will be considered in Chapter 5.

Terrorism by extremist political groups has now become a disturbing feature of everyday life and it has been suggested that such groups might be able to get their hands on nuclear material to make a crude explosive. This threat is certainly not underestimated and security on sensitive materials at nuclear establishments around the world and during transport between them, is extremely tight – probably at least as tight as the security on actual nuclear weapons already deployed in many countries of the world. Although it is not particularly reassuring, it is worth noting that there are easier ways of causing at least as much devastation as might result from a crude nuclear bomb and removal of one potential source of terror will, regrettably, do little to stop extremist groups finding others.

It is also worth stressing the difficulty of making a bomb if only to discourage the would-be terrorist from trying. It is true that the basic principles of a nuclear explosive can be found in any good encyclopaedia and a competent physics student should be able to produce a fairly detailed design on paper. But even within the military establishments of nuclear weapons countries it is doubtful whether there is one individual who could command all the technical skills involved in converting a paper design into an actual device. It needs a team with practical experience in a variety of specialist fields – such as chemical engineering, metallurgy, mechanical engineering, electronics, etc. – with well-equipped workshops to allow them to handle the nuclear material without contaminating themselves.

Is nuclear energy economical?

Operating nuclear power plants throughout the world are today producing electricity at a lower cost than alternative sources. The exact figures vary from country to country but typically the total cost of a unit of electricity produced from a nuclear power plant is about two-thirds the cost from a coal-fired plant and half the cost from an oil-fired plant.

A characteristics of nuclear electricity costs is that the majority is attributable to the initial capital cost of building the power station which typically is between 30 and 50 per cent higher than coal or oil plants. On the other hand the nuclear fuel costs are very much lower than coal and oil. This means that at least 60 per cent of the cost of a unit of electricity from a nuclear plant is the capital charge incurred in writing off the initial cost over a nominal life of between twenty and thirty years. As a result of inflation, early plants can maintain a competitive position with more efficient later plants because their initial capital cost was so much lower. If a plant exceeds its initial book life of twenty to thirty years – and there are several which will soon be doing that – their competitive position becomes even greater, due to the removal of capital charges. The opposite situation applies to fossil-fuelled plants where up to 80 per cent of the cost of a unit of electricity is attributable to the cost of coal or oil and rising fuel costs will have a greater impact on the older, less efficient plants than a new plant.

Because of the complex method of accounting for costs of electricity there is a great deal of scope for argument about comparative figures and hidden costs. It is sometimes suggested that use should be made of some form of inflation accounting in which the capital charges are increased with inflation to take account of the higher price that will eventually have to be paid for a replacement plant. At first sight this would appear to remove the apparent competitive advantage of early nuclear power plants. But to make a true comparison one would also have to apply inflation accounting to coal and oil production and make provision for the increasing costs of opening up new coal mines and oil wells in the present cost of these fuels.

It is also suggested that because the capital costs of nuclear power plants have tended to rise at a rate slightly faster than general inflation in recent years, and because of high interest rates, a new nuclear project undertaken today might not retain a competitive advantage over coal. The main reason for the rate of increase of capital cost is the demand for ever higher standards of safety and environmental protection, and to make a true comparison it is also necessary to assume higher standards of environmental control of pollution from coal-fired plants – in particular the provision of scrubbing systems to remove sulphur dioxide from the flue gases – which are likely to add substantially to the resulting electricity costs. Valid comparison is also complicated by the need to speculate on the likely cost of coal in about ten years' time when projects committed today would be ready to operate. On balance, careful estimates of future costs undertaken by reputable bodies in many countries suggest that nuclear power will maintain, or even increase, its competitive advantage.

Who do you believe?

The very vocal anti-nuclear movement which has grown up around the world over the past ten years would provide totally different answers to the questions posed in this chapter. Activists in this field appear to be of three kinds: there are local citizen groups genuinely concerned about any large-scale industrial development in their region; there are idealistic environmental groups who seek the post-industrial society and oppose nuclear energy as a symbol of the growth society; and there are activists of minority political parties from the right, left and centre who have embraced the anti-nuclear cause as an issue which will gain them greater public attention. Press attitudes towards nuclear energy have also become more hostile. In the 1950s and early 1960s there was an almost euphoric welcome for the wonders of the 'atomic age', but now the general enthusiasm for minority and anti-establishment causes has led many journalists into the nuclear energy debate. This in itself is no bad thing, but superficial and exaggerated reporting of what is undoubtedly a complicated subject has done a lot to stimulate fears among the public.

Another feature of the nuclear energy debate is that the anti-nuclear movement has produced its own experts, some of whom are highly qualified, others highly questionable. The nuclear industry believes that it has, time and time again, provided satisfactory answers to the technical questions raised by the anti-nuclear experts but it must be very difficult for the general public, with little understanding of the detailed topics raised, to decide which expert they should believe. The remaining chapters of this book aim to provide a basic guide to the workings of nuclear energy and to deal with some of the main technical issues. There are quite a few relatively complicated issues, but even members of the anti-nuclear movement would agree that the subject has many intriguing aspects which justify the effort to understand what it is all about. It has been said that a little knowledge is a dangerous thing. On the other hand, a little understanding of the basics of a subject can help in deciding which expert to believe.

2

Nuclear Reactors

The difference between the energy produced by burning familiar fuels and nuclear energy is related to the difference between the forces which hold atoms together and the forces which hold together the even smaller particles which make up the nucleus of an atom. If one visualizes these forces which bind the constituent parts of matter as powerful elastic bands wrapped around compressed springs it is not difficult to see that energy is released if the bands are cut and the springs fly apart. Much the same sort of thing happens if chemical bonds which hold atoms together in fuel molecules are broken or rearranged and if the nuclei of large atoms are split, or fissioned. The energy that is released is referred to as *binding energy* and the main difference between combustion and nuclear fission is that the binding forces which hold together the nucleus are very much stronger than chemical bonds. This means that it is more difficult to fission atomic nuclei but when it is achieved the energy released is a lot greater.

The binding energy released by making or breaking the chemical bonds of an individual molecule or the fissioning of a single nucleus is, of course, minute. To get useful power it is necessary to set off a self-sustaining chain reaction involving millions upon millions of atoms. This is what you are doing when you light a fire and feed it with fuel.

Fission chain reaction

It is only the largest naturally occurring atom, uranium, that is relatively easy to fission and even then only one type of uranium atom, or isotope, designated uranium-235, which constitutes about 0.7 per cent of naturally occurring uranium metal, is suitable. Very occasionally the large nucleus of such a uranium atom breaks up of its own accord or is said to undergo spontaneous fission. Among the fragments which fly off at great speed from a fissioning atom are two or three neutrons, one of two basic types of particle which make up the nucleus of all atoms. If one or more of these neutron fragments scores a direct hit on another uranium-235 nucleus it can break the bonds of that nucleus and cause a further fission. Two or three more neutron fragments will again be produced and these can cause further fissions.

This is clearly the basis of a chain reaction which could lead to the fissioning of millions upon millions of atoms and the release of useful energy. But it does not happen naturally in a lump of uranium metal for three main reasons: the spontaneous fission needed to start the chain reaction is a relatively infrequent event; the number of uranium-235 atoms in the lump of uranium will be much less than those of the predominant isotope, uranium-238, which is more likely to trap neutrons than be fissioned by them; and the neutron fragments tend to fly off from the fissioning atoms at such a high speed that the majority will escape from the lump of uranium without hitting any additional atoms of uranium-235.

One way to overcome the third of these problems is to find some way of slowing down, or moderating, the speed of the neutrons so that they have a greater chance of hitting more uranium-235 atoms. This can be done by interspersing lumps of uranium with a moderating material composed largely of small light atoms such as hydrogen or carbon (Figure 2.1). Neutrons tend to bounce from atom to atom in such moderating materials and in the process the speed at which they are flying around is reduced until it is comparable with the speed of motion of atoms in a hot gas. This is referred to as *reducing fast neutrons to a thermal state* and the result is that the lumps of uranium are permeated with a 'gas' of neutrons which then have a chance of causing further fissions.

Man first succeeded in producing a self-sustaining fission chain reaction in 1942 by inserting a large number of uranium rods into holes in a large pile of graphite blocks – graphite is a pure crystalline form of carbon. This experiment, conducted in a squash court in Chicago by a team led by the Italian scientist, Enrico Fermi, was referred to as an *atomic pile* but today it is more usual to refer to the assembly in which a self-sustaining chain reaction is produced as a *nuclear reactor*.

Another way of helping to sustain a chain reaction in a nuclear reactor is to increase the proportion of the uranium-235 isotope in the uranium. Because isotopes of a particular element are chemically indistinguishable, this is not an easy process but it can be done, by virtue of the small difference in weight between the two isotopes, in a physical process known as *enrichment*. As will be seen in Chapter 3, the enrichment process

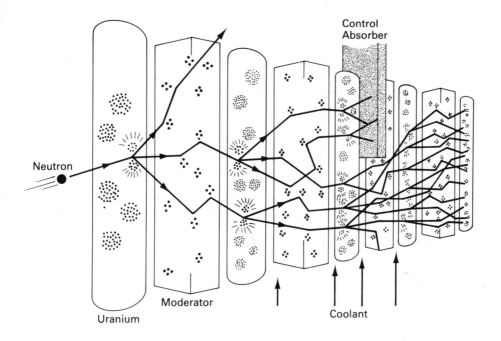

Figure 2.1 The fission chain reaction

is carried out in very extensive and costly plants but it is possible at reasonable cost to enrich the concentration of the uranium-235 isotope from the level of 0.7 per cent in natural uranium metal to between 3 and 4 per cent. In combination with a moderating material such as water, this allows a self-sustaining chain reaction to be produced in a reactor which is more compact than a reactor using natural uranium.

If the enrichment of the uranium is increased still further until the concentration of uranium-235 is over 20 per cent – or if the concentration of fissionable atoms is increased to a corresponding level by the addition of the man-made element, plutonium – it is possible to dispense with the moderator and get a self-sustaining reaction with fast neutrons. Such a reactor is known as a *fast reactor*.

Although it has been stated that a chain reaction does not start up of its own accord in a lump of natural uranium metal, it is interesting to record remarkable evidence that has been discovered to suggest that a natural chain reaction did in fact take place some 1.8 billion years ago in a uranium deposit at Oklo in the African Republic of Gabon. This is evidence that by introducing a moderating material and enriching uranium to make a reactor work, man has unwittingly imitated nature.

Nuclear Energy

Pockets of ground water would have provided the moderator for the Oklo reactor and the uranium in the rocks would have been 'enriched', with about 3 per cent uranium-235, compared with the 0.7 per cent of natural uranium today. This is because uranium atoms have, since the formation of the earth, been disintegrating very slowly by natural radioactive decay, with the uranium-235 isotope decaying at a faster rate than uranium-238.

Can nuclear reactors explode?

We know only too well that it is possible to make a nuclear explosive. But it is important to understand why, technically, this is very difficult to achieve and why it is impossible for civil nuclear reactors used for energy production to explode.

Anybody who has played the numbers game of repeatedly doubling or trebling a number will realize that the fission chain reaction, as described, has the potential to expand to huge proportions very rapidly. In practice the speed of the multiplication process is dependent on the number of neutrons produced by fissioning atoms that actually go on to fission further atoms. Many are captured by the more abundant atoms of uranium-238 or other materials which have to be used in the construction and cooling of a practical reactor; some neutrons escape from the bounds of the reactor; and some neutrons even disintegrate spontaneously into a proton and an electron. The chain reaction can only be sustained in what is referred to as a critical state, if there is a delicate balance in which, on average, one of the neutrons produced by the fission of one atom goes on to fission another atom and the remaining neutrons are captured or escape. A reactor can be controlled at this critical multiplication level of one by inserting into, or withdrawing from, the reactor neutron-absorbing materials which have a great affinity for capture of neutrons. If the neutron absorber, usually in the form of a control rod, is withdrawn the multiplication factor will rise above one and the number of fissioning atoms in the reactor will increase until the control rods are reinserted. This is referred to as *divergence* and is the means of increasing the power output from a reactor. If the control rods are inserted further the multiplication factor will drop below one and the number of fissioning atoms will decrease until eventually the chain reaction stops. This is referred to as the *shut-down*, or sometimes as *scram*, of a reactor.

When control rods are withdrawn from a reactor the power level only diverges relatively slowly. This is because not all of the neutrons which keep the chain reaction going are emitted instantaneously when an atom fissions. About 1 per cent only escape from the fission fragments after delays of up to a few minutes. It is the contribution of these delayed neutrons to the critical balance in a reactor which slows down the multiplication process when it is diverging and makes it easy to control.

Another important controlling factor in a reactor is provided by its response to an increase in temperature. This will provide a degree of self-regulation because, as the temperature rises, the materials in the reactor expand and the density of the fissionable atoms is reduced. This is usually enough to change the critical balance and reduce the multiplication factor to less than one. The power level and the temperature will then decrease.

To make a nuclear chain reaction run away at explosive speed it is necessary to achieve a multiplication factor well in excess of one so that the contribution of delayed neutrons to the critical balance becomes insignificant and the majority of the atoms are fissioned before they are blown apart. This can only be achieved with a very dense concentration of fissionable atoms and by minimizing the content of neutron-absorbing materials. To do this in an atomic bomb, a number of pieces of uranium metal in which the proportion of uranium-235 has been enriched to more than 90 per cent, or pieces of pure plutonium metal, are blown together with conventional explosives into a very dense critical mass.

In all types of practical reactor there are so many non-fissionable materials dispersed within and around the fissionable atoms that it is impossible to compress the fissionable atoms into a sufficiently dense pure mass to get the chain reaction to run away with explosive force. This is true even in the case of a fast reactor where the fissionable material is enriched to around 25 per cent and it is true even if the reactor gets so hot that it melts into a homogenous mass. At worst a reactor might get hot enough to cause a relatively low-powered explosion of a chemical nature in other materials within the reactor such as steam produced from cooling water.

Reactor types

There are a number of options for the component parts which make up a reactor. At the heart of the reactor is the fuel, which can be natural uranium if it is combined with an efficient moderator, or is more often slightly enriched uranium. It can take the form of metal rods but is more often formed into pellets – typically about 1 cm in diameter and 1.5 cm long – of uranium oxide, a ceramic material able to withstand high temperatures. Other fuel forms which have been considered and even demonstrated in small experimental reactors, range from liquid compounds of uranium to small particles of uranium carbide coated with graphite.

Nuclear Energy

The fuel is sealed into long tubes or cans of metal to form fuel elements and several thousand fuel elements dispersed throughout the moderating material will form the core of a reactor. The metal used to seal the fuel into elements is known as *cladding* and has to meet rather demanding requirements. It must be strong enough to retain the fuel and the smaller atoms which are formed when uranium atoms are split into two. These fission products are radioactive and it is therefore desirable to retain them within the fuel elements during the period that they are in the reactor. The cladding must, however, be relatively transparent to neutrons which have to be able to move freely between fuel elements to cause further fissions in the chain reaction. The energy released from the myriads of fissioning atoms causes the fuel to get very hot and the cladding material must be able to withstand this heat and at the same time allow the heat to radiate from the elements, rather like the radiation of heat from an element in an electric kettle.

A number of special metal alloys are used for cladding material in different types of reactor. These include stainless steel, which is an alloy mainly composed of iron, a magnesium-based alloy called Magnox, and a zirconium alloy called Zircaloy.

The heat radiated by fuel elements has to be got out of the core of a reactor if it is to be put to use. This is done with a gaseous or liquid fluid, sometimes referred to as a *heat transfer fluid* but more often known simply as a *coolant*, which is pumped past the fuel elements in the core of the reactor and carries the heat away to external heat exchangers. In addition to being an efficient heat transfer fluid the coolant must, like the cladding material, be relatively transparent to the neutrons passing to and fro between the fuel elements. Coolants that have been used in different types of reactor include: gases such as carbon dioxide or helium; water, either in its normal form, or heavy water which is enriched with the heavy isotope of hydrogen; metals such as sodium which melt at low temperatures and can therefore be maintained in a liquid state; and specially developed organic liquids.

The moderator, as previously mentioned, should be a material in which there is a predominance of hydrogen or carbon atoms because neutrons tend to bounce off these atoms rather than being captured. It happens, however, that the heavy isotope of hydrogen – deuterium – in which the nucleus is made up of a proton and a neutron instead of just a proton, is even less likely to capture neutrons and is therefore the most efficient moderator. Water, made up of hydrogen and oxygen, contains about 1 part in 5000 in which the molecules are made up with the deuterium isotope, and by separating this constituent one obtains heavy water which is a convenient, though somewhat expensive form of moderator. Ordinary water, referred to as light water, can also be used as a moderator but only in reactors with enriched uranium fuel, to overcome the slightly higher chance of capture of neutrons. Graphite is a pure crystalline form of carbon which is a sufficiently good moderator to be used in a reactor with natural uranium fuel as long as the fuel cladding and coolant are chosen from materials with good transparency to neutrons.

Different types of reactor design have been developed from different choices of the component materials. There is a large number of different possible combinations of the design options and each is likely to have some advantages and disadvantages. The evolution of any one type of reactor involves a very extensive and costly programme of research, development and testing and many prospective reactor designs have fallen by the wayside in this process. The most successful and widely used types of reactor fall into three main families:

Light water reactors – which use ordinary water for both coolant and moderator and therefore need slightly enriched fuel.

Gas-cooled reactors – moderated with graphite and using either natural uranium or slightly enriched uranium fuel.

Heavy water reactors – which use heavy water for the moderator and also, in most cases, for the coolant with natural uranium fuel.

These three types of reactor, all of which rely on the use of a moderator to reduce the speed of the neutrons in the chain reaction to thermal energies, are referred to generally as *thermal reactors*. A fourth family, which is not yet in widespread commercial use but which has been extensively developed because of its greater long-term potential, is the *fast reactor*.

Fast reactors – with no moderator, 20 per cent enriched fuel and, in most designs, liquid metal coolant (described in chapter 7).

Light water reactors

The most widely adopted type of reactor for nuclear power production around the world is the light water reactor (L.W.R.). The availability of enriched uranium from large plants built in the United States during the Second World War allowed the adoption of a reactor system using ordinary light water for coolant and moderator, both for military applications in submarine propulsion and in the development of large reactors for civilian power production. During the late 1950s and early 1960s a number of projects to demonstrate the capabilities of L.W.R. plants for electricity generation were put into operation in the United

States and, using American technology under licence, in a number of European countries and Japan. By the mid-1960s L.W.R. power stations were widely adopted for commercial electricity production in competition with coal- and oil-fired power stations.

To extract heat efficiently from a reactor it is necessary for the water coolant to be heated to temperatures around 300°C, well above the normal boiling point of water at 100°C. But the temperature at which water boils can be raised if it is kept at a high pressure just as it is in a domestic pressure cooker. A feature of light water reactors is the use of a massive steel pressure vessel to contain the reactor and pressurize the coolant. There are two types of L.W.R.: the pressurized water reactor (P.W.R.) in which the pressure is sufficiently high to maintain the coolant in a liquid state at all times, and the boiling water reactor (B.W.R.) where the pressure is such that there is controlled boiling of the water in the reactor.

Pressurized water reactor

The fuel for a P.W.R. consists of pellets of uranium oxide enriched to around 3 per cent and sealed into long tubes of Zircaloy to form elements about 3.5 m long and 1 cm in diameter. Bundles of elements are mounted together in open assemblies at the optimum spacing for the chain reaction in the fuel and surrounding water moderator. The assemblies usually have a square cross-section with, typically, up to 264 elements in each assembly. As many as 200 assemblies are loaded into a reactor for form a core with an approximate diameter of 3 to 3.5 m. This core of fuel assemblies is located on a grid support at the centre of the large steel pressure vessel.

The water coolant and moderator is forced to flow up through the fuel assemblies carrying the heat out through a number of heavy steel pipes around the top of the pressure vessel, Figure 2.2. Outside the vessel the hot water is circulated through steam generator units before being pumped back into the reactor vessel. In the steam generator the hot water from the reactor passes through a large bank of U-shaped steel tubes which are themselves surrounded by a vessel of water. The heat is transferred to the secondary water which, being at a lower pressure, boils to produce steam. This steam separates from the water at the top of the steam generator and is led off to drive a turbine and electrical generator. After giving up its energy in the turbine the steam is condensed to water and pumped back to the bottom of the steam generator.

Depending on the power output of a reactor there will be between two and four primary circuits, or loops, connected to the pressure vessel,

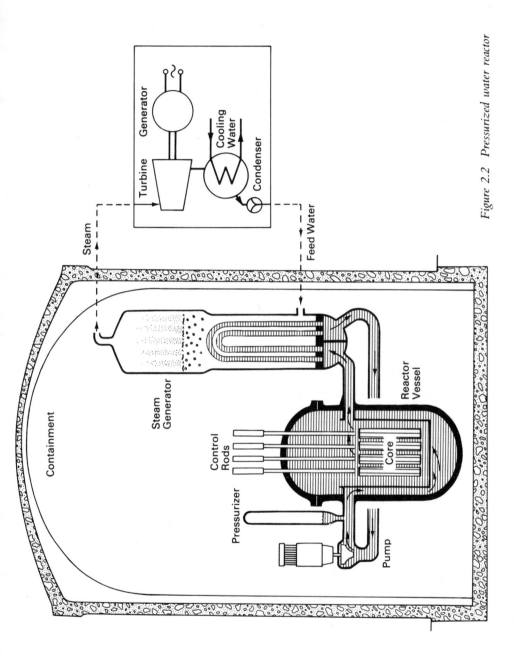

Figure 2.2 Pressurized water reactor

each supplying one steam generator. One of these primary loops will also have a connection to an additional vessel which is used to pressurize the whole system to about 130 times normal atmospheric pressure (2000 lb/sq in or 160 kg/sq cm).

Neutron-absorbing control rods, similar in size to the fuel elements, can be inserted into a number of spaces within each fuel assembly. The reactor is started up by withdrawing the control rods but they are held up by electromagnetic latches which, if de-energized for any reason, allow the control rods to drop back into the core and shut down the reactor. Slower control of small fluctuations in the operating power level of the reactor can be achieved by adding small quantities of a neutron-absorbing liquid – boric acid – to the water passing through the core of the reactor.

The useful life of fuel assemblies in the core of a P.W.R. is about three years. During this time there is a slow depletion of the uranium-235 atoms as they are fissioned – referred to as *burn-up* – and a build-up of the smaller fission product atoms within the fuel. Some of the fission product atoms are strong absorbers of neutrons and can therefore affect the critical balance of the chain reaction. It is the eventual build-up of the number of these absorber atoms to a level where the chain reaction cannot be maintained that determines the useful life of the fuel and not the complete burn-up of the uranium-235. In practice, about one third of the fuel assemblies in the core of the reactor are changed at intervals of one year to eighteen months. This is done with the reactor shut down. The top of the pressure vessel, which is held on with massive bolts, is taken off in a deep pool of water above the reactor and the fuel assemblies are moved using long handling tools. The thickness of the pool water provides protection from radiation.

An important feature of the P.W.R. is that it is a compact and robust plant. This feature dictated its early choice for submarine propulsion and has had an important impact on the economics of the subsequent commercial nuclear power plants. The major component parts, such as the pressure vessel, the steam generators and the circulating pumps can be manufactured in factories using well-established engineering techniques and this makes it easier to impose the very high standards of quality control demanded in the nuclear industry.

The compactness of the core also means that a lot of heat is being produced in a relatively small volume – referred to as a high power density – and it therefore becomes important to maintain the flow of high pressure water coolant to remove the heat. This need to maintain coolant flow continues for some time after the reactor has been shut down because of decay heat generated by the radioactive fission products in the fuel. For this reason P.W.R. plants are provided with a number of back-up cooling systems. These are designed, in particular, to deal with the remote possibility of a serious failure of the highly pressurized primary coolant circuit. Any loss of coolant would result in partial depressurization of the system with the likelihood of boiling of the water and reduced effectiveness of the coolant in the reactor core. The emergency core cooling systems are therefore designed to inject cold water at high pressure into the core region in the event of a circuit failure.

Other safety aspects will be considered in more detail in Chapter 4 but at this stage it is worth drawing attention to the multiplicity of barriers which, as with all reactor systems, are designed to prevent the escape of the radioactive fission products that accumulate in the reactor fuel. First, the ceramic nature of the uranium oxide fuel pellets tends to retain most of the fission product atoms. Then the cladding material provides a barrier of high integrity. The water coolant is a partial barrier in that any escaping fission product atoms tend to dissolve in the high temperature water. The massive steel pressure vessel and the pipework of the primary circuit are designed and manufactured to very high standards to provide a further barrier. The reactor vessel and primary pipework are situated in concrete vaults of massive construction which, although not designed to be leakproof, would still retain a large proportion of any escaping material. Finally, the whole plant is surrounded with a large containment building – often double-walled – which is designed to withstand the steam pressure which would result from a massive rupture in the primary coolant pipework.

Boiling water reactors

In early consideration of the likely consequences of loss of pressure and boiling of the water coolant in an L.W.R. it was recognized that it would be possible to establish stable boiling conditions in which there is a high degree of self-regulation. This is because the water in the core is also acting as a moderator for the chain reaction. If the core starts to get too hot more steam bubbles are formed in the water and this reduces the amount of moderator around the fuel, which slows down the reaction and reduces the temperature.

Boiling water is also an effective way of removing heat from the core of a reactor and transferring it to a point of use because, as it boils, water absorbs extra heat – known as *latent heat* – and when the steam is subsequently condensed the latent heat is released. Early B.W.R.s were similar to P.W.R.s, with a primary circuit transferring the hot steam to a heat exchanger where steam was generated in a secondary circuit to drive a steam turbine and generator. But in later commercial designs of B.W.R.s great simplification was achieved by taking the steam directly from the top of the reactor vessel to the turbogenerator and then, after condensing the steam, recirculating the water back to the reactor vessel (Figure 2.3).

In practice it is also necessary to circulate water within the reactor vessel – up through the core region and down around the outside of the core. The steam produced by boiling in the core is separated from the remaining water by swirling vanes in a space at the top of the reactor which throw the water out to the side for recirculation and allow the steam to rise to the top of the vessel from where it is piped off to the turbogenerator. The circulation of water within the vessel is effected with pumping systems situated around the reactor core.

B.W.R. fuel assemblies are very similar to those of P.W.R., with clusters of long elements containing enriched uranium oxide pellets sealed in Zircaloy cladding tubes. The spacing of the elements within the assemblies is slightly larger, allowing for the reduced moderator efficiency of the boiling water surrounding them. As a result the diameter of the core is larger, typically around 4 m and, together with the need for a water recirculation system around the outside of the core and the space above the core for steam separation, this calls for a larger size of pressure vessel. On the other hand the operating pressure for the boiling water conditions is about half that of the P.W.R. – around sixty times atmospheric pressure – and therefore the pressure vessel does not need to be of quite such heavy construction.

The control rods in a B.W.R. are normally in the form of cruciform plates of neutron-absorbing material which are inserted into the space between the fuel assemblies. Because of the steam separation above the core the control rods are inserted from below the core. Although this means that you cannot rely on gravity to insert the control rods, it has been possible to design a very reliable hydraulic system which will fire the rods into the core to scram the reactor. The reactor power output can also be controlled by varying the speed of circulation of water through the core because this, like the self-regulation mechanism already mentioned, changes the proportions of water and steam bubbles in the space between the fuel elements and therefore controls the chain reaction.

The simplification achieved by direct use of steam from a B.W.R. to drive a turbogenerator, together with the self-regulating characteristics, are important features of the B.W.R. Plants of this type were the first to break into the truly commercial market in the United States in competition with the then cheap electricity from oil-fired plants, and have retained about a one third share of that market since. Outside the United

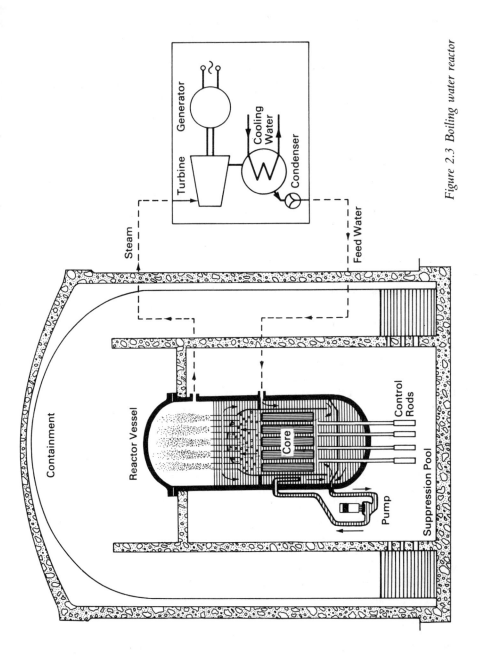

Figure 2.3 Boiling water reactor

States the B.W.R. has found favour in Sweden and Japan but in most other markets, for commercial rather than technical reasons, the P.W.R. has tended to dominate.

At first sight it seems that the B.W.R., which does not use a heat exchanger to separate the primary coolant circuit from the secondary steam-raising circuit, has one less barrier against the escape of radio-activity than the P.W.R. However, the water circulation within the reactor vessel would retain a large proportion of any radioactive material that might escape through failure of fuel – element cladding and the pipes carrying steam from the reactor to the turbogenerator can be shut off by fast-acting valves to isolate any release within the primary system. A further protective barrier is also provided in the form of a pressure suppression system connected to the concrete vault around the reactor vessel and the main pipes. Any blow-out of steam would be directed by large ducts into a deep pool of cold water where it would be condensed harmlessly. This also reduces the pressure that would have to be withstood by the containment building around the whole plant in the unlikely event of a large burst in the primary circuit.

Gas-cooled reactors

The first countries to adopt nuclear power on a sizable scale for commercial electricity production were Britain and France in the late 1950s. At that time neither country had access to supplies of enriched uranium but they had developed natural uranium-fuelled reactors, with graphite moderators and gas cooling, for military production of plutonium. It was therefore this type of reactor that was the only available system for the state-owned utilities to use for electricity production. Later both countries independently developed the technology for enrichment of uranium but while Britain went on to develop a more advanced gas-cooled reactor using enriched uranium fuel the French switched to the use of the P.W.R. The first generation gas-cooled reactors are usually known as Magnox reactors, taking their name from the magnesium alloy used for the fuel cladding, and the later design is known simply as the Advanced Gas-cooled Reactor (A.G.R.). A great deal of development work has also been undertaken in Europe and the United States on an even more advanced gas-cooled reactor which operates at a very high temperature and is known as the High Temperature Gas-cooled Reactor (H.T.G.R. or H.T.R.). This system will be described in Chapter 7 as one of several advanced reactors which offer considerable promise for the future but which have not yet been introduced on a commercial scale.

Magnox reactor

The fuel in a Magnox reactor consists of rods of natural uranium metal about 1 m in length and 3 cm in diameter, sealed in a can of Magnox cladding to form fuel elements. These elements are loaded one on top of another in vertical channels in a massive pile of graphite moderator blocks, and carbon dioxide gas coolant is forced up through the channels and around the fuel elements. The pile of graphite blocks which incorporates the core of a typical Magnox reactor is about 14 m in diameter and around 8 m high and will contain around 30 000 fuel elements stacked in several thousand channels. To improve the effectiveness of the gas coolant in removing the heat from the fuel elements the gas is maintained at a high pressure – about nineteen times atmospheric pressure (270 lb/sq in or 19.5 kg/sq cm). At this pressure the gas has more the consistency of a liquid.

In the early designs of Magnox reactors the large reactor core is contained inside a huge steel pressure vessel (Figure 2.4). Gas pumps, usually referred to as *blowers*, circulate the high pressure carbon dioxide through the core region and out through large ducts to a number of heat exchanger units where steam is produced in large banks of tubing. Because of the similarity of these heat exchangers to conventional boilers in which steam is produced in banks of tubes heated with the hot gases from the burning of coal or oil, they are often referred to as *nuclear boilers*. The steam produced drives a turbogenerator to produce electricity in the normal way.

In later designs of Magnox reactors, large pre-stressed concrete pressure vessels have been adopted. In these the strength is provided by steel cables stretched to a high tension and embedded in massive concrete walls. A central cavity lined with steel and insulation to protect the concrete from the hot gas, contains the reactor core and the banks of boiler tubes situated around the core region. Gas blowers projecting into the bottom of the vessel circulate the coolant up through the core and down around the outside through the boilers.

Control rods are inserted into additional channels in the pile of graphite moderator blocks. They are raised and lowered by chain drive mechanisms with a magnetic clutch which allows them to be released and fall by gravity into the core region to shut down the reactor rapidly.

A characteristic of a reactor fuelled with natural uranium is the need to change the fuel relatively frequently because of the build-up of fission products in the fuel which absorb neutrons. Due to the low concentration of uranium-235 in natural uranium, these fission product absorbers start to have an impact on the critical balance of the chain

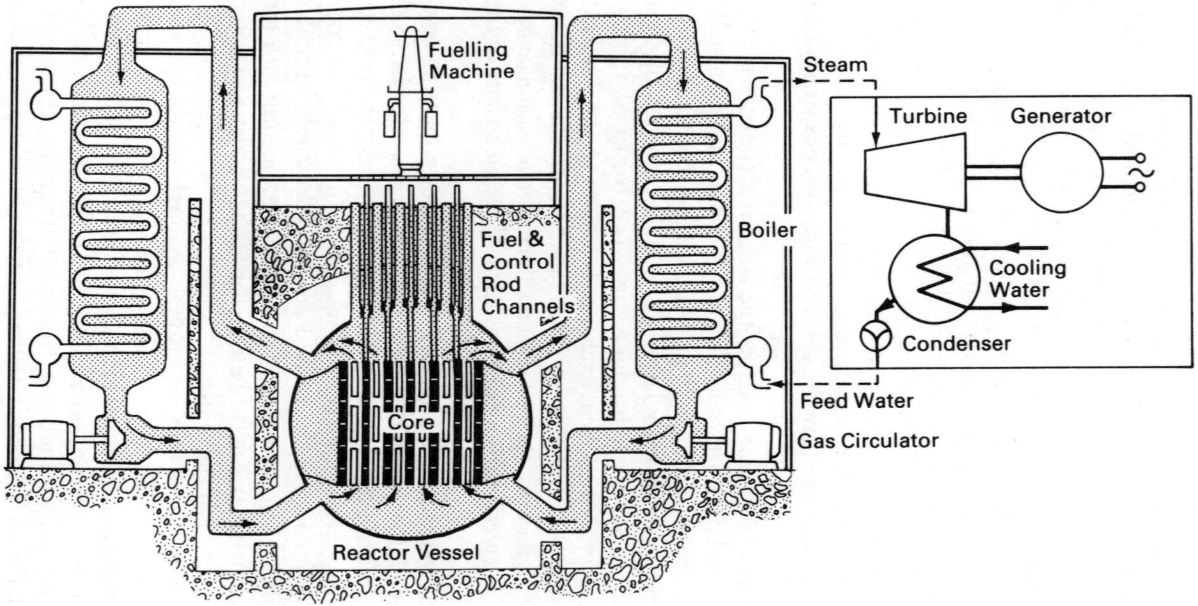

Figure 2.4 Magnox gas-cooled reactor

reaction at a lower level of fuel burn-up than is the case in reactors with enriched uranium fuel. Machines have therefore been developed to charge and discharge the fuel elements in individual channels of the reactor while it is operating. These fuel-handling machines are connected to penetrations through the top of the pressure vessel. With remote grabs they are able to draw a string of fuel elements out of a channel and up into a heavily shielded chamber in the machine on top of the reactor. Because of the tens of thousands of fuel elements in the core of the reactor, fuelling operations of this kind do not significantly disturb the power of the reactor while they are being carried out.

Although it is necessary to maintain the gas coolant at a high pressure in order to increase the efficiency of heat removal from the reactor core, partial loss of pressure due to an improbable failure of the gas circuit is not as critical as it is in the case of a water-cooled reactor. This is partly because there is no sudden change from liquid to gaseous state and also because of the larger reactor core with a lower power density. In addition the graphite moderator has a large capacity for absorption of heat and, as long as the reactor has been shut down, it would take many hours for the residual heat from the radioactive decay of fission products in the fuel to raise the temperature to dangerous levels, even with the coolant at atmospheric pressure.

Magnox reactors were expensive to build, partly because of their sheer size and also the resulting need to carry out a large proportion of fabrication work at the power station construction site. When first built it was expected that the capital cost of the power plants spread over the life of the plants would make the cost of electricity slightly higher than that from coal- and oil-fired power stations, but in the event coal and oil prices have risen so fast that the Magnox stations now produce extremely competitive electricity. But while the existing Magnox power stations will continue to make a very valuable contribution to electricity supplies for many years, no serious consideration is given to building any more reactors of this type because of the massive inflation of capital costs since the late 1950s and early 1960s and the fact that more efficient systems offering lower capital costs are now available.

Advanced gas-cooled reactor

The main objective for improvements in a second generation of advanced gas-cooled reactors (A.G.R.s) was to achieve a higher operating temperature. This allows the generation of 'dry' steam comparable to that produced in conventional fossil-fired boilers and improves the overall efficiency – the ratio of the electrical power generated to the heat produced in the reactor core – from around 30 per cent in a Magnox reactor to 50 per cent in an A.G.R.

The main limitation on the operating temperature is the need to ensure an adequate safety margin between the operating temperature and the melting point of the fuel-cladding material. In improving on the Magnox alloy, consideration was initially given to the use of the metal beryllium which is even more transparent to the neutrons that keep the chain reaction going and has a higher melting point than Magnox. Manufacture on a commercial scale, however, proved to be too difficult and instead stainless steel was adopted. This has the desired high melting point but absorbs more neutrons than Magnox. To compensate for the greater neutron absorption in the stainless steel it was therefore necessary to adopt enriched uranium fuel.

The fuel in an A.G.R. is in the form of uranium oxide pellets, similar to those used in light water reactors, with the uranium-235 content enriched to between 2 and 2.5 per cent of the uranium. The pellets are sealed into stainless steel cladding tubes to make elements 1 m long and 1.5 cm in diameter, and thirty-six of these elements are mounted in a cylindrical sleeve of graphite to form a fuel assembly. About 2500 of these fuel assemblies are loaded into channels in a large pile of graphite moderator blocks similar to those in a Magnox reactor but, by virtue of the enriched uranium fuel, the overall size of the reactor core is somewhat smaller – around 8 m high and 10 m in diameter.

The A.G.R. is contained in a large pre-stressed concrete pressure vessel similar to the later Magnox reactors (Figure 2.5) although to increase further the efficiency of heat removal from the core the carbon dioxide coolant gas is maintained at a higher pressure – around 40 times atmospheric pressure (580 lb/sq in or 42 kg/sq cm). Other features are similar to the later Magnox reactors: the coolant is circulated by blowers projecting into the bottom of the vessel and passes up through the fuel channels and down through the boiler tubes arranged around the core. Steam produced in the boiler tubes drives the turbogenerator to produce electricity.

Like the Magnox reactor, the A.G.R. core temperature would rise relatively slowly if there was an escape of coolant gas and partial loss of pressure but because of the higher operating temperature and more compact core some back-up cooling systems are provided to deal with unlikely emergencies. At the same time, the use of a ceramic uranium oxide pellet and stainless steel cladding provide more secure primary barriers against the escape of radioactive fission products from the fuel.

The first five A.G.R. power stations to be built in Britain ran into a

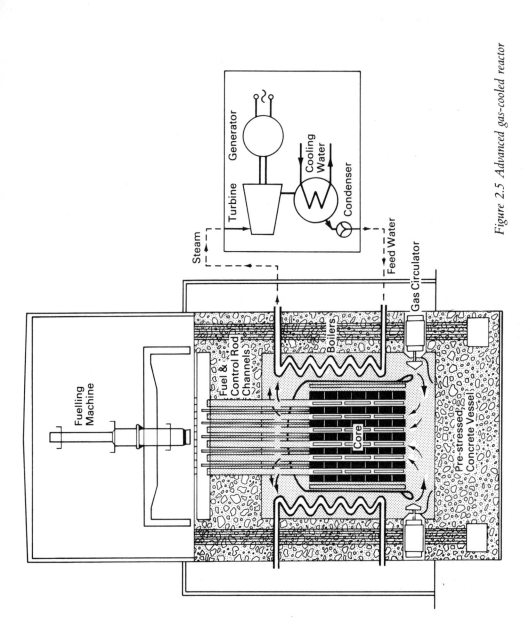

Figure 2.5 Advanced gas-cooled reactor

number of technical problems and suffered considerable delays in construction, largely due to failure to take full account of the engineering implications of a substantial increase in the operating temperature and pressure of the gas coolant. Early operation of the first of these plants is starting to confirm their high efficiency and it is expected that at the end of the day they will produce electricity which is very competitive with that from coal- and oil-fired power stations. But the relatively large size and the high proportion of site construction rather than factory prefabrication in the building of A.G.R. power stations means that they are always likely to be more expensive than light water reactors. This fact, together with the much more successful and widespread commercial development of light water reactors, means that the use of A.G.R.s has been confined to Britain, where two more stations are being built, but a switch to the P.W.R. for the later part of the 1980s is under consideration.

Heavy water reactors

A number of different reactor concepts have been developed in different parts of the world to make use of the very good moderating characteristics of heavy water. The majority of these reactor designs use a large tank of heavy water with a number of tabular channels passing through it to hold the fuel assemblies and carry a flow of coolant. Where they differ is in the choice of coolant. Prototype heavy water reactors have been built with high pressure carbon dioxide as a coolant, with organic liquids which do not need to be pressurized and with light water maintained at the appropriate pressure for controlled boiling. But only one reactor design, using high pressure heavy water as a coolant, has been developed to the stage of large-scale commercial exploitation in Canada. Known as Candu, this reactor has also been sold to customers in India, Pakistan, South Korea, Argentina and Romania.

Candu reactor

The fuel in a Candu reactor consists of natural uranium in the form of ceramic pellets of uranium oxide. The fuel pellets are sealed in relatively short Zircaloy cladding tubes – about 500 cm long – to form elements, and fuel assemblies are made up from bundles of up to thirty-seven elements. About 4500 of these elements are loaded into 380 fuel channels which pass horizontally through a cylindrical tank of heavy water moderator about 6 m in length and diameter (Figure 2.6). Heavy water coolant is circulated past the fuel in zirconium pressure tubes passing through each of the fuel channels in the tank of moderator. In this way

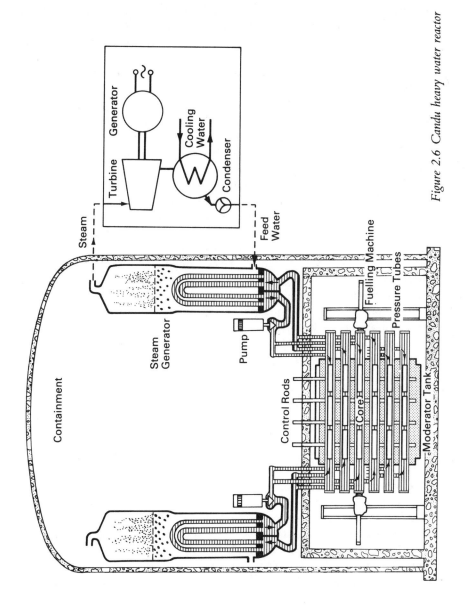

Figure 2.6 Candu heavy water reactor

Nuclear Energy

the heavy water coolant can be pressurized to limit boiling in a large number of relatively small tubes – about 100 mm in diameter – instead of a massive pressure vessel. The moderator is insulated from the hot coolant by the gap between pressure tubes and their surrounding channels in the moderator tank. As a result it can be kept at a low temperature and does not need to be pressurized.

A network of pipes connects the pressure tubes to a primary coolant circuit similar to that in a pressurized water reactor. The coolant is circulated by pumps through banks of U-tubes in eight steam generator units with ordinary light water being boiled on the secondary side to produce steam and drive a turbogenerator.

The main control of a Candu reactor is provided with neutron-absorbing control rods which can be inserted into vertical channels which also pass through the large tank of heavy water moderator. In early designs provision was also made for shut-down of the reactor by rapid draining of the heavy water moderator into a dump tank below the reactor. In more recent designs a greater diversity of control systems is used, including an arrangement for injecting neutron-absorbing liquids into the moderator. Fuel is loaded into and removed from the reactor while it is operating, using remotely-operated machines which can connect to both ends of individual pressure tubes. New fuel assemblies are inserted at one end and spent fuel assemblies are drawn out at the other end into a heavily shielded flask.

Emergency systems for high pressure injection of cooling water into the primary coolant circuit are provided to cope with loss of heavy water coolant through a breach in the pipework. The whole of the reactor plant is contained in a large concrete containment building which is also provided with a means for reducing the pressure of any steam that might escape from a major pipe failure. In Candu power stations with several adjacent reactor units this pressure supression system has been provided by connecting the containment buildings by large ducts to a huge common vacuum chamber. If there were to be a sudden excess pressure in any one of the containment buildings it would blow open vents in the connecting ducts and the steam would be sucked into the vacuum chamber. An alternative arrangement which has been adopted in single reactor units is to provide a large reservoir of water in the roof of the containment building which can be released into the building as a fine dousing spray to condense any escaping steam and so reduce any excess pressure.

Production of heavy water is a difficult and expensive process but after the initial filling of the moderator tank and the primary coolant circuit there is only a requirement for relatively small amounts to make good losses during the operation of the plant. For this reason the initial capital cost of a Candu reactor is higher than that of a light water reactor. On the other hand there is no need for enrichment of the uranium fuel – a process which is also difficult and expensive – and therefore the Candu reactor enjoys lower fuel costs during its operating life and on balance is competitive with light water reactors. Candu reactors at power stations in Canada have performed very well and the system has attracted interest in a number of developing countries of the world where there is a desire to avoid dependence on other countries for supplies of enriched uranium fuel.

3

Nuclear Fuel

The fuel for a nuclear reactor is not just dug out of the ground and burnt like coal but is subjected to a number of careful refining processes and is fabricated with great precision into fuel elements and assemblies before it is loaded into a reactor. After use the spent fuel may be subjected to further reprocessing so that radioactive waste products can be separated safely and unused fuel or by-product plutonium can be recycled to fuel further reactors. This sequence of processes is referred to as the *nuclear fuel cycle*.

There are a number of different options that can be adopted within the nuclear fuel cycle and the subject has in recent years assumed political significance because of the suggestion that certain fuel cycles might be less vulnerable to clandestine diversion of nuclear materials for military purposes. An exhaustive international study in which some forty-six nations and five international organizations participated was completed in 1980. Known as the International Nuclear Fuel Cycle Evaluation (I.N.F.C.E.), it concluded that none of the different fuel cycles was completely resistant to the risk of abuse but at the same time none was incompatible with a non-proliferation regime – that is an international regime of controls to prevent proliferation of nuclear weapons. These political aspects of nuclear power will be discussed further in Chapter 5 but here we will describe the steps in the basic fuel cycle proposed at the outset of civil nuclear power development.

The fuel cycle

To understand the flow of material around the basic fuel cycle it is worth imagining a loop of pipes in which the size of the pipe represents the volume of the material flowing through it (Figure 3.1), though in prac-

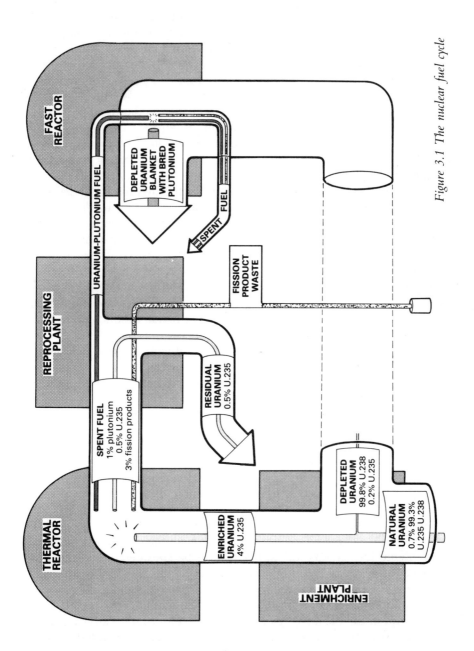

Figure 3.1 The nuclear fuel cycle

tice the fuel cycle is broken down into a lot of individual steps. As described in the previous chapter, naturally occurring uranium is composed of two main isotopes – uranium-238 and uranium-235 – in the proportion of 99.3 per cent to 0.7 per cent. Although the two isotopes are mixed together indistinguishably they are represented diagrammatically as a large pipe carrying uranium-238 with a much smaller pipe inside it carrying uranium-235. The process of enrichment can then be represented as splitting the flow into two, with a larger proportion – 3 to 4 per cent – of the uranium-235 in the enriched uranium product and a smaller proportion – about 0.2 to 0.3 per cent – in a reject stream known as depleted uranium.

When enriched uranium fuel is fed into a thermal reactor – that is a reactor in which the neutrons in the chain reaction are moderated to thermal energies – it is part of the uranium-235 flow which is 'burnt' (fissioned) to produce energy. At the same time some of the uranium-238 is converted by neutron absorption in the reactor into a different atom – plutonium. The spent fuel coming out of the reactor can therefore be represented still as a large pipe carrying mainly uranium-238 but now with three constituents in smaller pipes – about 0.5 per cent of unburnt uranium-235, some 3 per cent of radioactive fission product atoms and about 1 per cent of plutonium.

In a reprocessing plant these different constituents of spent fuel can be separated chemically. The radioactive fission products are extracted as a small volume of highly concentrated waste material. The bulk of the uranium containing unburnt uranium-235 can be recycled to the enrichment plant where it will provide some new fuel for a thermal reactor. A smaller amount of uranium, together with the plutonium at a concentration of around 40 per cent, is a suitable fuel for use in a fast reactor. In this case some of the plutonium is 'burnt' and the spent fuel contains fission product wastes, unburnt plutonium and new plutonium which can be separated and recycled by passing the spent fuel back to a reprocessing plant.

At the same time the compact core of the fast reactor can be surrounded by some of the depleted uranium left over from the enrichment process and which, for the sake of simplicity, can be considered as consisting entirely of uranium-238. Neutrons escaping from the core of the fast reactor will be absorbed in the blanket of uranium-238 and in the process more plutonium will be produced. It is actually possible to produce plutonium in the blanket region at a slightly faster rate than it is being burnt in the core of the reactor. This plutonium can be separated in a reprocessing plant to refuel the fast reactor with surplus plutonium, building up over a number of cycles until there is enough to fuel another fast reactor. Thus, after initial feed of fast reactors with plutonium produced as a by-product from thermal reactors, the fast reactor can become self-supporting except for the feed of depleted uranium which would otherwise have been a waste product.

This process is referred to as *breeding* and the fast reactor is often called a fast breeder reactor. The importance of breeding is in the fact that, over a period, it allows the bulk of the uranium-238 in the original natural uranium to be converted into a useful fuel and at the end of the day it is possible to derive about fifty to sixty times more energy from a given quantity of uranium ore than would be obtained if it was used only to fuel thermal reactors.

The alternative of cutting the fuel cycle short before reprocessing and treating all of the spent fuel and the depleted uranium from the enrichment process as waste products is referred to as the *once-through fuel cycle*. Although a much smaller fraction of the natural uranium is converted into useful energy this approach can be attractive from an economic point of view because it is still possible to get 10 000 or more times as much energy from a ton of uranium as one gets from a ton of coal or oil. And as will be seen later, reprocessing plants and fast reactors are very expensive to build.

In thermal reactors such as the heavy water reactor or the Magnox gas-cooled reactor it is, as already explained, possible to achieve a fission chain reaction with natural uranium fuel and there is no need for the enrichment plant in the fuel cycle. In this case it is still possible to reprocess the spent fuel to separate plutonium and provide the initial feed for fast reactors.

Yet another fuel cycle option is to reprocess spent fuel from a thermal reactor and recycle the plutonium as an alternative to uranium-235 enrichment in the thermal reactor. This is known as *thermal recycle* and allows one to obtain about 30 to 40 per cent more energy in a light water reactor from an original quantity of uranium ore.

Finally, there are a number of alternative fuel cycle options, involving the breeding of nuclear fuel from naturally occurring thorium metal instead of depleted uranium. In this case the thorium can be converted, either in the blanket of a fast reactor or in a thermal reactor, into another isotope of uranium – uranium-233 – which is a very good nuclear fuel and can be substituted for uranium-235 enrichment in any type of reactor.

Although the different fuel cycle options offer many interesting possibilities it is important to recognize that the establishment of all the industrial processes which make up the different steps in the cycle involves many years of expensive development work. Each step is in effect an individual service industry involving complicated technical and economic problems.

Uranium resources

Although deposits of uranium are found in most regions of the world and uranium also occurs in very low concentrations in sea water, the world resources for the purposes of nuclear energy are related to the cost of mining and extraction and also depend on whether the uranium is used with relatively low efficiency in today's thermal reactors or whether it is eventually used as feed material for fast breeder reactors. To date the largest known resources which can be mined at economic prices for once-through use in thermal reactors are in the United States, due in part to favourable geology but also to the fact that exploration for uranium in that country has been very extensive. There are also large proven reserves of uranium in Canada, Australia and the African continent. It is reasonable to assume that the large land areas of the Soviet Union and China have large, economically attractive deposits of uranium but no reliable figures are available concerning the extent of resources or level of production. From a general knowledge of the geology of South America it is also to be expected that there are good prospects for finding uranium but exploration to date has yielded only moderate finds.

Significantly, Western Europe and Japan, with large programmes for the use of nuclear power, have very little uranium which could be mined at attractive prices for use in thermal reactors. There are a number of small deposits in France which are being exploited and are capable of meeting about half the country's present requirements and there are large quantities in Sweden which occur in a low grade form in shale deposits but which are not economical to mine at current world prices. It is for this reason that Western Europe and Japan are particularly interested in maintaining development of the fast breeder reactor which uses the uranium fifty to sixty times more efficiently than a thermal reactor and could therefore be run on higher cost uranium if world market conditions or political constraints become more difficult in the future.

At present, however, known resources of low cost uranium in the non-Communist world are fairly plentiful at just under 2 million tonnes and estimates of additional resources which are considered to be likely to exist would add a further 1.5 million tonnes. If the price that consumers are prepared to pay for uranium were to double then the firm resources would be increased by a further 0.75 million tonnes, and estimated additional resources could be increased by a further 1 million tonnes. Intelligent speculation, based on general geological considerations, sug-

Nuclear Fuel

gests that further uranium resources that might, or might not, be discovered in the future are in the range of 6.5 to 15 million tonnes.

To put these uranium resources into perspective, and taking a figure of around 5 million tonnes as being a realistic estimate of the amount of uranium recoverable at reasonable costs in the non-Communist world, it would take over 60 billion tonnes of coal or around 40 billion tonnes of oil to produce the same amount of electricity as would be obtained by using the uranium in thermal reactors. That is about one tenth of the recoverable coal resources or one-fifth of the ultimately recoverable oil resources. If, however, this amount of uranium is eventually used in fast reactors it would be equivalent to around 10 trillion tonnes of coal or about eighteen times the recoverable coal resources of the non-Communist world.

For practical planning, however, whether one is talking of uranium, coal or oil, the rate at which the resources can be produced is quite as important as the ultimately recoverable resources. With emerging new industries such as uranium mining and nuclear power generation it is extremely difficult to make accurate projections of supply and demand. The situation is made more complicated by the impact of alternative reactor strategies and fuel cycles and the political controversy which surrounds decisions on nuclear power programmes in many countries. The great range of uncertainty is illustrated in Figure 3.2 which is based on a periodic review of uranium resources, production and demand undertaken by the O.E.C.D. Nuclear Energy Agency and the International Atomic Energy Agency. It shows the demand for uranium based on high and low projections of expansion of nuclear power programmes in the non-Communist world. There is a large divergence of possible demand after the year 2000, depending on whether large-scale introduction of fast reactors takes place after the mid-1990s or, alternatively, whether there is continued reliance on thermal reactors only. Also shown is the maximum attainable rate of production from existing known resources of uranium which is likely to be ahead of demand during the 1980s but could start to decline in the 1990s unless some of the speculative resources are converted into real finds of uranium by exploration during the next ten years. It can be seen that if there is a slow rate of expansion of nuclear power programmes, coupled with large scale introduction of fast reactors at the end of the 1990s, uranium production from existing resources should be able to keep pace with demand. But at the opposite extreme, if there is rapid expansion of nuclear power programmes and continued reliance on thermal reactors, a shortage could develop after the turn of the century unless there are substantial new discoveries of uranium.

Both the mining industry and the nuclear power industry hope for an

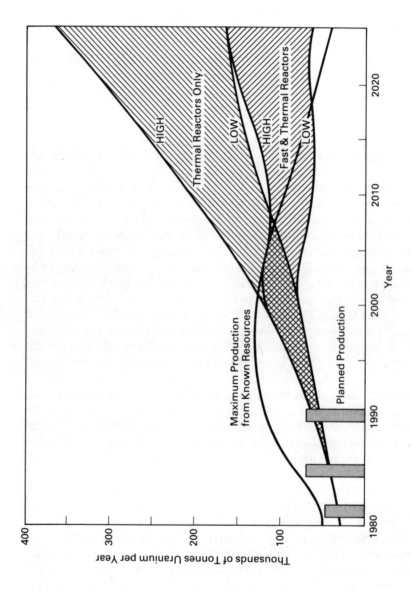

Figure 3.2 Uranium demand and supply projections

orderly development somewhere between these two extremes. But to achieve orderly development will require timely decisions: on investment in exploration; on the opening up of new mines, which can take eight to ten years; on the construction of nuclear power stations, which can take around ten years; and on commercial introduction of fast reactors, which could be spread over twenty years or more.

Uranium mining

Most large-scale production of uranium today comes from rock formations close to the surface of the earth which can be extracted by conventional open cast mining techniques. Underground mining is also involved in some areas where the uranium occurs in vein-type deposits and may become more important in the future as the more accessible surface deposits are worked out. In all cases the actual concentration of uranium within the rocks is small. Large deposits near the surface can be commercially viable with average uranium concentrations as low as 0.02 weight per cent. More typically, much of the commercial mining takes place in areas where the uranium concentrations are in the range 0.1 to 0.4 weight per cent and, apart from a few isolated occurrences of high concentrations of uranium, the richest known ore bodies only have average grades around 3 weight per cent.

In spite of the low concentration of uranium in rocks, exploration for surface deposits is fairly easy because of the natural radioactivity of uranium and its radioactive decay products. Large-scale initial surveys are usually carried out from the air using sensitive radiation measuring instruments. Promising areas are then followed up with more precise surface measurements of radioactive emission but final assessment of the commercial prospects of deposits can only be made with extensive programmes of drilling into the rock formations.

Mills at, or close to, the uranium mines are essential to extract the uranium from the bulk of the rock. The rocks are first crushed and the metallic constituents are dissolved out with acid. Chemical processes known as *solvent extraction* or *ion exchange* are then used to produce a relatively pure compound of uranium called yellow cake. It is in this form that shipment of natural uranium normally takes place.

Uranium is also found in association with other valuable materials, notably gold in South Africa, which improves the economics of extraction even with very low grade uranium ores. In many regions of the world low grade uranium ores have also been found in association with phosphate rocks and there is the potential for extraction as a by-product of phosphate fertilizer production. The chemical extraction processes are, however, quite difficult and there is a need to invest in a large-scale fertilizer production operation before the by-product uranium becomes economically competitive at present world market prices.

Like all mining operations the extraction of uranium-bearing rocks involves occupational hazards but the hard crystalline structure of the rocks involved, or large-scale mechanization which is possible with open cast deposits, reduces the risks of normal accidents as compared with deep coal mining. The direct radiation emission from the uranium represents an insignificant health hazard because the concentration of uranium is very small and the nature of the radiation is such that most of it is absorbed within the surrounding rock. There is, however, a gaseous radioactive decay product of uranium, radon, which normally would also be trapped within the rock but can be released during the course of mining operations. In the early rush of uranium mining to meet military requirements there were instances of miners in underground workings being exposed to levels of radiation from radon which would not be considered acceptable today and there is evidence that these workers suffered a higher than normal incidence of cancer in later life. Improved ventilation in underground mines provides a simple method of removing the released radon, and in all present-day operations radiation exposure levels to miners are kept well below the internationally accepted limits. Certainly the occupational health hazards in uranium mining are very much less than the risk of lung diseases from the dust in coal mining.

Mill tailings, in the form of fine sand-like material produced from crushed rock after the uranium has been removed, will still contain the radioactive daughter products which constitute most of the radioactivity associated with naturally occurring uranium. If the tailings are allowed to settle back into the ground or are covered with a layer of top soil, there is no reason to believe that the residual radioactivity will be any more mobile than it was in the original uranium deposit. There were some early cases of tailings from uranium mines being used for what appeared to be a very good material for the preparation of foundations for buildings, only to cause concern later due to the slow build-up of the radon gas in the buildings, but this too can be dealt with by improved ventilation of the buildings.

As with all mining operations there are continuing efforts to improve health and safety standards in the production of uranium. But because it is possible to produce such a lot of energy from a small amount of uranium, the occupational dangers involved in uranium mining are very much smaller than the cost in human lives involved in other more conventional sources of energy. Even allowing for the fact that it may be necessary to mine 1000 tonnes of rock to produce 1 tonne of uranium, there is still a need for at least a tenfold increase in the scale of mining operations to produce an equivalent amount of energy from coal.

Uranium is usually purchased from the mining companies by the eventual operators of nuclear power plants on long-term contracts, but before it is ready to be fabricated into fuel it has to be subjected to further refining processes. If the uranium is used in the natural uranium form in the early gas-cooled reactors or heavy water reactors, the additional processes merely consist of chemical conversion to high purity metal or oxide prior to fabrication of the fuel. If, as is now the more general case, the uranium is to be enriched, there is a need to convert it first into a gaseous form that is suitable for feeding into an enrichment plant. Uranium hexafluoride – UF_6 – is a chemical compound which is gaseous at normal temperatures and therefore suitable for enrichment purposes. The conversion of yellow cake to uranium hexafluoride involves specialized chemical processes which constitute another important commercial service in the fuel cycle. At present these conversion services are carried out for the most part in the industrialized countries where they have been developed as part of a large nuclear power programme, but in the longer term there are moves by some of the larger uranium mining countries to refine their product to a higher degree, and therefore a higher value, before exporting it, and conversion to uranium hexafluoride is likely to be the first activity to be undertaken in pursuing this general objective.

Uranium enrichment

Different isotopes of an element can have markedly different properties in a nuclear reactor because of the different numbers of particles which make up their nuclei, but because the chemical properties are governed mainly by the electrons around the nucleus, the different isotopes of an element are virtually identical chemically. This makes it very difficult to devise large-scale processes to separate isotopes. The processes that are available rely on the small difference in weight between the isotopes and, in the case of uranium where the isotopes of interest have weights of 235 and 238 units, this difference is a small fraction of the total weight. This means that uranium enrichment is a particularly difficult process and, although the principles of various separation methods are well known, only a few countries have developed the technology for large-scale commercial plants.

The main process which has been used to date is called *gaseous diffusion* and was first developed in the United States during the Second World

War as part of the Manhattan project. It relies upon the fact that if a gas diffuses through a porous material, lighter atoms will pass through more easily than heavy atoms. If uranium hexafluoride gas is compressed on one side of a membrane containing very small holes, then the gas which passes through to the other side of the membrane will have a very slightly higher concentration of the lighter uranium-235 and the gas that stays on the first side will be very slightly depleted in uranium-235. If this separation is repeated many thousands of times it is possible to obtain a product gas stream in which the concentration of uranium-235 has been enriched from the natural level of 0.71 per cent to around 3 or 4 per cent while the by-product stream of gas has the uranium-235 concentration depleted to around 0.25 per cent. It takes many more thousands of stages of separation to produce a highly enriched product with around 90 per cent uranium-235 which is required for military purposes.

In practice the separation membranes are arranged in the form of banks of tubes so that a large surface area can be exposed to a flow of compressed gas on one side and the product gas can be extracted from the other side. The bank of tubes is contained in a steel vessel connected to an electrically driven gas compressor. The compressed gas is also passed over a bank of tubes containing cooling water to remove the heat generated when the gas is compressed. The slightly enriched gas collected from the product side of the membrane tubes is piped to the compressor of another unit while the depleted gas stream is passed back to a previous unit. In this way the many thousands of separation stages are connected together with an increasingly enriched product progressing in one direction and a gradually depleted by-product moving in the opposite direction.

In a typical commercial plant, one tonne of natural uranium feed introduced in the middle of the line of separation stages would produce a fifth of a tonne of enriched uranium product at the top end of the line and four-fifths of a tonne of depleted uranium at the other end. It can be seen therefore that the actual flow of material at the two ends of the line of separation stages is less than it is at the feed point in the middle and therefore the size of the units can be progressively reduced. In practice there are usually about three sizes of separation units in a typical installation and the line of optimally sized units is referred to as a *cascade*.

Gaseous diffusion enrichment plants consume a large amount of electricity in driving the thousands of compressors but the enriched uranium product allows nuclear reactors to generate electricity more efficiently. As a result it is possible to visualize the enrichment process as a method of effectively storing the electrical energy consumed in the process in the form of a more concentrated nuclear fuel. Gaseous diffusion plants are also strongly influenced by economy of scale. A very large plant supplying a large market requires access to a low cost supply of electricity. This was the situation with the large gaseous diffusion plants built initially for military purposes in the United States and subsequently converted to supply the civil market. More recently a large commercial plant has been brought into operation in France and in this case the source of cheap electricity is a nuclear power station with four 900 MWe pressurized water reactors built on a site adjacent to the enrichment plant. At full capacity the enrichment plant will use about three quarters of the output from the nuclear power station but in the process it will supply the uranium enrichment needs of a further eighty to ninety reactors of comparable size.

The second important technique for the separation of isotopes makes use of very high speed centrifuges. By spinning the centrifuge at high speed heavier isotopes are thrown outwards more vigorously by centrifugal force and with suitable design of a collection system it is possible to get a partial separation into slightly enriched and slightly depleted streams. When this method was first looked at as a possible method for separating isotopes of uranium it was found that the speed of rotation that would be needed to provide a reasonable degree of separation was higher than could be achieved by centrifuges at that time. More recently, the availability of high strength lightweight alloys, and even the possibility of centrifuges reinforced with carbon fibre, revived interest in this method as a contender for commercial enrichment, especially as it offered the possibility of lower electricity consumption than gaseous diffusion plants.

A joint centrifuge venture was started by Britain, Germany and the Netherlands in 1970 and plants are now operating commercially in Britain and the Netherlands with another being built in Germany (see Plate 1). A feature of the centrifuge plant is that an enrichment cascade is built up from many thousands of identical machines and increased throughput can be achieved by increasing the number of cascades in parallel. Another important variable in the centrifuge plant is the length of the cascade. Increasing the length of the cascade with more stages produces higher enrichment. Unlike gaseous diffusion plants where the economy of scale demands a commitment to a very large enrichment plant and subsequent economics of operation requires that this large block of capacity is filled with long-term contracts, the centrifuge plants can start small and grow in step with the market demand for enrichment services.

The latest addition to commercial enrichment capacity in the United States will also be making use of centrifuges. A factor here is the tenfold reduction in electricity compared with gaseous diffusion plants and the fact that there is no longer a ready availability of cheap electricity as there was when the large military gaseous diffusion plants were converted to civil use. The Japanese have also developed centrifuge enrichment technology to the stage of a pilot scale plant and plan eventually to build their own commercial plant.

Many other processes for uranium enrichment have been examined, and are re-examined from time to time, but only two have been developed to the stage of small-scale pilot plants. These are known as aerodynamic processes because they rely upon the swirling of a stream of gas without the need for high speed spinning machinery such as the centrifuge. In a German development known as the *jet nozzle technique* a flow of uranium hexafluoride gas mixed with helium is forced at very high speed through a narrow slit and over a curved surface. Centrifugal forces again cause a partial separation of heavier atoms as they fly around the corner and a suitably placed knife edge splits the gas flow into a slightly enriched and slightly depleted stream. The other aerodynamic process has been developed in South Africa and makes use of a system called a vortex tube. In this the gas, again a mixture of uranium hexafluoride and helium, is forced through a vortex aperture which causes it to swirl at high speed down the length of a tube, achieving the same effect as a centrifuge without rotation of the walls of the tube. The electricity needed to drive the aerodynamic processes falls between that of gaseous diffusion and centrifuges and it will be necessary to make a commitment to a relatively large plant in order to achieve competitive commercial operation.

A newer enrichment technique which is being developed in the United States makes use of laser energy to ionize uranium atoms in such a way that the heavy and light isotopes can be separated electromagnetically. As a late starter it is unlikly that this technique will be adopted for large-scale commercial enrichment services in the near future but it could find an application in the United States to strip more enriched uranium from large stockpiles of depleted uranium left over from earlier military enrichment when in the interests of throughput the depleted stream was rejected with a higher concentration of residual uranium-235 than is now adopted in commerciacial practice.

Finally, the French have been working on an alternative chemical enrichment technique. Chemical exchange processes have been known for some time, in which the transition from one chemical form to another is accompanied by a slight priority for lighter or heavier isotopes even though their chemical properties are almost identical. They were generally excluded from consideration as contenders for commercial enrichment services because it is necessary to have a very large inventory of uranium in the process plant to achieve a reasonable rate of output of enriched product. However, this very feature means that it would take an extremely long time to produce significant quantities of highly enriched uranium of a grade suitable for the construction of a nuclear explosive. The chemical process is therefore being developed to satisfy a political requirement for a process which could be offered to other countries, especially those with sizable reserves of uranium, for enrichment to the low levels required for nuclear power plants without incurring the risk of proliferation of nuclear weapons production.

The output of an enrichment plant is specified in separative work units – or S.W.U. This is a measure of the amount of processing carried out to produce the required level of enrichment and is the basis on which customers are charged for the service. To provide the initial load of fuel for a typical 1000 MWe light water reactor would require about 50 tonnes of uranium enriched to around 3 per cent. To produce this it would be necessary to feed about 250 tonnes of natural uranium into the enrichment plant and subject it to 200 000 S.W.U. Thereafter the annual refuelling requirements of the reactor would be about half these quantities.

The three large enrichment plants in the United States which have been converted from military to civil production have a total potential capacity of nearly 27 million S.W.U. per year and further capacity of around 9 million S.W.U. per year will be provided with a new centrifuge plant in the second half of the 1980s. The large new gaseous diffusion plant in France will be operating at a capacity of 10.8 million S.W.U. per year in the early 1980s. The centrifuge plants in Britain, the Netherlands, and later Germany will expand their combined capacity from 600 000 S.W.U. to between 2 and 10 million S.W.U. per year during the 1980s according to demand for enrichment services. The only other major supplier of commercial enrichment services during the 1980s will be the Soviet Union which has contracts for the supply of around 2 million S.W.U. per year to Western customers, but there is no knowledge of the total available capacity. At present the total availability of enrichment services appears to be more than enough to meet the demands of projected nuclear power programmes during the 1980s but during this period decisions will have to be taken on the building of new plants to meet anticipated growth in demand in the following decade.

Heavy water production

Reactors fuelled with natural uranium do not have a need for enrichment of uranium in their supporting fuel cycle but if they are moderated and cooled with heavy water there is a requirement for a comparable service which can be described as enrichment of water. The heavy water moderator and coolant of a reactor are, of course, different from the fuel in that they are not consumed during operation of the reactor apart from some inevitable small losses and degradation. There is therefore a requirement for a large initial inventory – typically about 250 tonnes for the moderator and 170 tonnes to fill the coolant circuit in a 500 MWe reactor – and then only a small continuing demand to make good losses.

Hydrogen, at the opposite end of the Periodic Table from uranium, is the smallest and lightest atom with just one proton in the nucleus of the predominant isotope. The heavy isotope of hydrogen has one proton and one neutron in its nucleus and is therefore approximately twice the weight of the main isotope. The fact that there is this much larger relative difference in the weight of hydrogen isotopes compared with the isotopes of uranium means that the processes for separation are somewhat easier. On the other hand the concentration of molecules of water containing the heavy hydrogen isotope is only 0.02 per cent in naturally occurring water and it is necessary to produce 99.8 per cent pure heavy water for use in reactors. There is therefore a need to process huge quantities of water to obtain the required tonnages of heavy water.

In the early days of nuclear research, electrolysis of water was used as the method of separating heavy water. The ions of hydrogen travel slightly more easily to the electrodes than ions of heavy hydrogen and therefore the residual water after electrolysis is partially enriched with heavy water. Repeating the process many times will yield almost pure heavy water but the process consumes large quantities of electricity and was only possible in locations where there was an abundant source of cheap hydroelectric power.

For large-scale commercial production of heavy water more recent plants make use of a chemical exchange process in which water is brought into contact with hydrogen sulphide gas. In this process heavy hydrogen atoms tend to migrate from the water molecules to gas molecules at high temperatures and in the opposite direction at low temperatures. The process takes place in a series of very tall vessels in which, initially, water cascades down from the high-temperature top end over a series of trays, and hydrogen sulphide passes up from the cool bottom end. Over two stages this will produce a 30 per cent concentration of the heavy hydrogen atoms in the gas stream taken off from the middle of the vessels and a reverse process is then used in further vessels to transfer the heavy hydrogen back to heavy water molecules. Further upgrading to 99.8 per cent can then be achieved by repeated distillation in which the light water tends to boil off first.

Nuclear Fuel

Canada has a number of commercial heavy water plants with a total production capacity of over 4000 tonnes per year. This is more than enough to meet the needs of the domestic nuclear power programme and also supply heavy water for reactors that have been sold to other countries. The Indians, who originally bought heavy water reactors from Canada, have developed independent capability in this field including the production of heavy water, and other countries adopting heavy water reactors are likely to follow a similar path since one of the reasons for choosing this type of reactor in the first place was to achieve greater self-sufficiency than is possible with reactors dependent on foreign fuel enrichment services.

Fuel fabrication

The assemblies of fuel elements spend up to three years in the core of a power reactor. They generate high temperatures and are subjected to a powerful flow of coolant fluid, usually at high pressure and often quite corrosive. In addition there is very intense nuclear radiation associated with the fission chain reaction. And the spacing of the long thin elements must be maintained with great precision throughout its period in the reactor. For these reasons fuel assemblies must be designed and manufactured to the very highest engineering standards.

Straightaway it should be said that the critical importance of the design and quality of manufacture of fuel elements has been fully appreciated from the very outset of nuclear power development work and a tremendous effort has been devoted to the production of fuel for each type of nuclear reactor. In particular every design of fuel assembly is subjected to the most rigorous testing in research reactors, often under simulated conditions which are more severe than will be encountered in power producing reactors. Largely as a result of this great concentration of early development work on the fuel assemblies, the experience universally in commercial power stations has for the most part been exceptionally good and most operators have found that the performance of the fuel has exceeded the designers' expectations.

Fuel assemblies for nuclear power plants are fabricated in purpose-built modern factories. Because of the fundamental characteristic of nuclear energy in which a very great deal of energy is derived from a very small weight of fuel, a modest-sized factory can manufacture enough fuel to supply the annual needs of ten to twenty large nuclear power plants.

The fabrication plants are divided into two main areas. In the first the uranium is prepared in the right chemical and physical form for loading into cladding tubes to make elements, and in the second the elements are fabricated into fuel assemblies. In most types of reactor, as has already been seen, the fuel is in the form of small cylindrical pellets of uranium oxide, typically about 13 mm long and 8 mm in diameter. To produce the pellets the uranium arriving from the enrichment plant in the form of uranium hexafluoride gas in special containers, is first converted to uranium oxide powder. Great care is taken to achieve a very consistent powder and to control very rigidly the purity. The powder is then pressed with a binding compound into the pellet shape. These pellets, which are still rather soft and crumble easily, are then heated in a high temperature furnace to produce a very hard sintered pellet with a ceramic consistency rather like black unglazed pottery. The surfaces of the pellets are usually ground to produce very precise final dimensions and they are carefully examined for any imperfections. Any defective pellets are crushed to powder again and recycled to the start of the pellet production process.

The final operation in the first section of the fuel fabrication plant is to pack and seal the pellets into tight fitting tubes of cladding material to form fuel elements. The tubing, of very high quality, is supplied by well-established tube manufacturers but most of them have had to establish a special tube-making plant to meet the tight specifications and extreme cleanliness demanded by the nuclear industry. Even so the fuel fabricator will still subject all the tubing to further inspection before using it. When filled with pellets and a small spring to keep them packed together an end cap is welded onto the cladding tube to complete the sealed element. Each element is subjected to careful leak testing and radiographic examination to ensure the integrity of the welds and the correct packing of the pellets inside.

Throughout the first section of the plant there is not only very tight quality control on the precise composition of the fuel material but also very careful accounting for the movement of all material through the plant. This is partly because the uranium is very valuable but also because such plants are subjected to stringent control to ensure that no significant amounts of uranium can be diverted from the plant for unauthorized purposes. When the elements pass through into the second section of the fabrication plant they have a specific identity and are accompanied by a large batch of paperwork giving precise details of the batch of uranium used and all the associated quality control results.

In the second section of the plant the operations are mainly precision machining and assembly, still under conditions of extreme cleanliness and subjected to very tight quality control. The elements are mounted in assemblies with support grids spaced along their length to ensure correct

Nuclear Fuel

spacing. End fittings provide for precise location of the assemblies in the core of the reactor and also for connection to the special handling equipment used to load them into and remove them from the reactor.

Large private companies involved in the development and supply of reactors, especially in the United States, have established their own impressive plants for fuel fabrication and these are usually the preferred suppliers of fuel, at least for the first few core loadings of a new reactor. But independent fuel fabrication companies have also been set up in a number of countries and have shown themselves able to compete and meet the very demanding quality requirements while supplying replacement fuel for many different reactors. The nature of the work – high quality processing and engineering in relatively small modern factories – has made fuel fabrication a service activity of early interest in countries with developing nuclear power programmes even if it does initially involve the establishment of licensing arrangements with the major manufacturing organizations in the United States and Europe. Some smaller industrial countries with traditional skills in high quality engineering have also shown interest in this specialist field of activity. Belgium, for example, established an impressive capability in fuel fabrication and in one plant, now jointly owned and financed by French and Belgian commercial interests, has manufactured the bulk of the first fuel loadings for the large number of pressurized water reactors being put into service in France. (See Plate 2.)

Other countries, notably Britain, developed the technology of fuel fabrication within the national research organizations, along with the other service activities in the fuel cycle. All these commercial activities have now been split off from the research organizations with the establishment of a state-owned fuel cycle company.

Spent fuel storage and transport

When fuel assemblies are taken out of a nuclear reactor after a period of up to three years producing power they are referred to as *spent fuel* or *irradiated fuel*. Although the content of radioactive fission products in the spent fuel is only a few per cent by weight they emit very intense radiation and continue to generate a significant amount of heat. For this reason the fuel assemblies must be handled remotely and provided with some form of cooling. But this residual heat and radiation emission falls off rapidly with time. One minute after the reactor is shut down the heat output falls to about 3 per cent of the operating level, after an hour it is down to 1 per cent and after one day it is about 0.5 per cent. In the case of light water reactors it is likely to be at least a week after shut-down before the lid of the pressure vessel is lifted to allow removal of the fuel and by this time the heat level will be down to about 0.3 per cent of the operating level. But even at these levels direct exposure to the radiation from the spent fuel would be very dangerous and it is therefore stored for a further cooling-off period in a deep pool of water alongside the reactor. After a year of cooling off the residual heat would be down to 0.07 per cent and after ten years 0.04 per cent.

In the case of gas-cooled reactors, the cladding material of the fuel elements was not designed for extended immersion in water and the cooling-off period is therefore limited by consideration of corrosion. This problem is most severe in the case of the Magnox-clad natural uranium metal fuel elements and after a period of between six and eighteen months it is desirable to transport this spent fuel in heavily shielded containers to a reprocessing plant for further treatment. With water-cooled reactors, both heavy and light water, the Zircaloy fuel element cladding is designed for severe operating conditions in high pressure hot water and there is no reason to believe that the spent fuel could not be stored for almost indefinite periods in the far less severe conditions of cooling pools. The period of storage is then dictated by the capacity of the pools at the reactor site and political considerations about the timing and desirability of reprocessing, which will be considered in Chapter 5.

Most light water reactors around the world have been designed with capacity in their cooling pools for about five years spent fuel discharge but the introduction of special compact storage racks can extend this period to around ten years. Pending decisions on reprocessing or the availability of reprocessing capacity it is clear that in the next twenty years there is going to be a build-up of spent fuel in excess of the storage capacity at reactor sites and in some countries, notably Germany and Japan, this problem has already become a source of some embarrassment to power station operators. Plans are therefore in hand to provide an extra step in the fuel cycle: the interim storage of spent fuel in large central facilities with capacity to accept the spent fuel from a large number of reactors. In addition, the large reprocessing plants being built in Britain and France will be provided with large reception pools so that they can provide storage of spent fuel for up to ten years before it is reprocessed.

Transport of spent fuel from reactors to either reprocessing plants or interim storage facilities is carried out in massive steel containers – usually referred to as *flasks* – which are designed and tested to very stringent international standards. Testing of these containers includes a drop test with the most vulnerable corner impacting on a solid unyielding surface and a ferocious fire test. In the United States dramatic demonstrations have also been carried out with transporters driving at full speed into concrete walls and with a locomotive driving into the side of a transporter carrying a container. Tests in Germany include the firing of a projectile at the side of a container to simulate the crashing of a supersonic jet fighter. In all these tests the containers have survived without any loss of leak tightness but even if there were small leaks the possibility of escape of dangerous amounts of radioactivity from within the fuel is very remote, because the elements are still clad in a material which has been designed to withstand the far more severe temperature and pressure conditions within an operating nuclear reactor.

Reprocessing

Well known and relatively simple chemical processes are used to separate radioactive fission products from unused uranium and plutonium in spent fuel. The basic principles of these reprocessing techniques were made known at the first International Conference on the Peaceful Uses of Atomic Energy in Geneva in 1955 and from the earliest days of the development of nuclear power for civil purposes it was assumed that spent fuel from nuclear power plants would, sooner or later, have to be reprocessed. Reprocessing allows the concentration of the radioactive waste products of fission into the smallest volume and opens the way to the recycling both of unused uranium and the new fuel, plutonium. It is against this background that one should view the development of reprocessing activities to the scale of an important commercial service industry. The more recent questioning of the desirability and need for reprocessing dates only from recent political concerns about nuclear weapons proliferation, raised in 1977 and discussed in Chapter 5.

The chemical process used in reprocessing is known as *solvent extraction*. It involves the preferential transfer of certain chemical elements from water-based – or aqueous – solutions to solutions in organic liquids, and subsequent separation of the aqueous and organic liquids by a simple settling process, as with oil and water. The aqueous solution of the spent fuel, complete with uranium, plutonium and all the fission products, is produced by dissolving the spent fuel in very strong nitric acid. The aqueous solution is then mixed with an organic solvent and most of the radioactive fission products stay in the aqueous solution while the uranium and plutonium transfer preferentially to the organic ultimate solution. The units in which this process takes place are known as *mixer-settlers* although in more recent plants they have been replaced by *pulsed columns* in which there is a counter-current flow of the aqueous and organic solvents over pulsating baffles. Two stages of processing in this way are enough to strip more than 99.96 per cent of the fission product wastes from the uranium and plutonium. The uranium and plutonium are then back-washed into the water. With slight adjustment of the aqueous solution to change the chemical valency state of plutonium it is possible in a further stage of solvent extraction to obtain a separation of the uranium and plutonium products. Further processing of the different streams is carried out to purify the uranium and plutonium products.

While the basic chemistry of reprocessing is relatively simple it is complicated in practice, particularly in the initial stages, by the high level of radioactivity. It is therefore necessary to design plant of very high reliability for remote operation within heavily shielded concrete cells. As the uranium and plutonium are separated in pure form in aqueous solutions where the water can act as a moderator, it is also important to avoid concentrations which could reach the critical conditions for a fission chain reaction. The know-how for the building of a commercial reprocessing plant is, therefore, not so much concerned with the basic chemistry but rather with safe, reliable and economical solutions to these practical problems.

The practical problems which are likely to have the most influence on the smooth throughput of a reprocessing plant are associated with the section of the plant known as the *head end*. This is where the fuel material is separated from the cladding and dissolved in nitric acid. With the uranium metal fuel elements of the early gas-cooled reactors the relatively soft Magnox cladding is stripped off, rather like peel off a banana, by remotely-operated machines in heavily shielded cells or at the bottom of a deep pool of water. Head end decladding operations of this kind have been carried out on a routine basis for more than twenty years at reprocessing plants in Britain and France and generally the experience has been good. But the need for periodic maintenance of the decladding machines has on occasions caused a temporary bottleneck in reprocessing activities. Since the Magnox-clad spent fuel has only a limited life in storage pools it is important to avoid bottlenecks and for this reason new head end facilities are being built in Britain to provide greater spare capacity and, using the fund of past experience, to provide more reliable and easily maintainable decladding machines. Once the cladding from the natural uranium metal fuel elements has been removed, batch dissolving in nitric acid and feeding of the solution into the main chemical separation plant has proved to be relatively straightforward.

With uranium oxide fuel of the kind used in light water reactors and advanced gas-cooled reactors the Zircaloy or stainless steel cladding cannot be stripped off mechanically. Instead the assemblies of fuel elements are chopped into small pieces with powerful hydraulically-operated shearing machines. The small pieces of fuel elements, about 30 mm in length, are then put into nitric acid and the oxide fuel is leached out from inside the cladding tube. Clearly the remotely-operated shearing machines are also a potential source of bottlenecks but techniques are being perfected in a number of small-scale plants around the world, and the ability to store spent fuel for long periods reduces the pressure to maintain a steady flow of material through the head end plant.

Dissolving of oxide fuel and the initial stages of chemical separation have, however, proved to be rather more difficult than originally expected. This is because, with the higher burn-up used with oxide fuel, the dissolution is not as complete as it is with metal fuel, and particulate matter has to be separated from the solution in centrifuges before it is fed to the main chemical separation process. Another factor which has to be taken into account in the reprocessing of enriched uranium oxide fuel is the fact that each element will have spent an appreciably longer time producing more power in a reactor than natural uranium fuel. As a result the concentration of radioactive fission products will be greater – though still about the same in proportion to the amount of electricity produced – and it is necessary to take great care to avoid degrading of the organic solvents, due to exposure to intense radiation, in the first stage of chemical separation. There is also a higher concentration of residual enriched uranium and plutonium and greater care has to be used in designing the plant so that there is no possibility for a build-up of sufficient volume to provide the critical conditions for a fission chain reaction.

Many years of commercial operation of reprocessing plants for natural uranium fuel have been accumulated at large plants in Britain and France. In the 1960s it also looked as if plans for the provision of commercial reprocessing of enriched uranium oxide fuel in the United States, Europe and Japan would provide a capacity in excess of the needs for the 1970s. In the event a combination of early technical problems, political constraints and the fact that the need to reprocess oxide fuel is not so pressing, has resulted in only very small quantities of spent fuel being reprocessed in this period.

The present situation is that the United States has a virtually complete commercial reprocessing plant with a capacity to handle around 1500 tonnes of spent fuel per year which is prevented from operating for domestic political reasons. France is starting to fulfil commercial reprocessing contracts using an oxide head end with a potential capacity of 400 tonnes per year feeding into one of their existing natural uranium reprocessing plants. During the 1980s this capacity will be doubled and two new reprocessing plants of 800 tonnes per year capacity are to be purpose-built for oxide fuel. In Britain, early operation of an oxide head end feeding into the existing natural uranium reprocessing plant ran into technical difficulties and priority is now being given to the building of a large new oxide reprocessing plant which will provide a capacity of 600 tonnes per year in the second half of the 1980s. Germany has been operating a pilot scale reprocessing plant with a capacity of 40 tonnes per year off and on throughout the 1970s but plans to build a large commercial plant have been temporarily blocked for political reasons and current plans are for one or more 350 tonnes per year plants, designated as demonstration projects, to operate in the late 1980s. Japan has been operating a demonstration reprocessing plant with a potential capacity of 200 tonnes per year since the end of the 1970s but plans for a large commercial plant are unlikely to be realized before the early 1990s.

Fuel re-fabrication

To complete the recycle of unused uranium and plutonium fuel separated from fission product wastes in a reprocessing plant it is necessary to make new fuel assemblies, either for use in the same type of thermal reactor or for the fuel of fast reactors. In the case of thermal reactors the unused uranium can be fed back to the enrichment plant to supplement new uranium and in this case it becomes indistinguishable in the fabrication of new fuel. But a more interesting possibility is to provide effective enrichment of the uranium with plutonium either to a level of around 3 per cent for thermal reactors – referred to as *thermal recycle* – or to a level of around 20 per cent for more efficient use in fast reactors.

To make this fuel uranium and plutonium are chemically converted into the form of oxide powders and then carefully mixed together in the appropriate proportions. After this the production of sintered pellets and fabrication of fuel assemblies is basically the same as the original fuel fabrication except that it has to be carried out in glove boxes or in completely automatic plants, mainly because of alpha radiation emission from the plutonium. Alpha radiation has very low penetration and can be stopped with a good pair of rubber gloves but until the fuel pellets are inside the cladding tubes great care has to be taken to avoid direct contact or inhalation of plutonium oxide dust. An alternative process involving the so-called gel precipitation of plutonium oxide in the form of microspheres offers the prospect of less dust and is likely to be adopted in future manufacture of mixed oxide fuel.

So far the need for mixed oxide fuels has been confined to the production of relatively small batches of fuel for test use in thermal reactors and fuel for experimental and prototype fast reactors. A great deal of development work has, however, been carried out on re-fabrication techniques in a number of countries, and small-scale production plants have been established in Britain, France, Belgium, Germany, Italy, Japan and the United States. Large commercial plants for fuel re-fabrication will follow the introduction of fast reactors on a larger scale or if political decisions are taken to recycle plutonium in thermal reactors.

Waste management

Strictly speaking, the safe management and disposal of radioactive waste materials is the final stage in the fuel cycle. But because this topic impacts not only on the various stages of the fuel cycle but also on routine operations at nuclear power plants, research establishments, hospitals and other industrial users of radioactive isotopes, and also because it has become a subject of particular public concern, it will be considered in a special chapter. It is worth mentioning here, however, that viewed as a commercial service activity the scale and cost of waste management activities are relatively modest compared with the other fuel cycle services. A number of companies, as well as national nuclear research centres, offer efficient processes for volume reduction and fixing of low level wastes and the costing for eventual solidification of high level wastes is included within reprocessing contracts.

Repositories for final disposal of high level wastes will in themselves be fairly large projects, but because they will serve the lifetime requirements of all the power stations in a large industrial country the shared costs will be small and should be adequately met by funds already being set aside by the operators of nuclear power plants in most countries.

4

Nuclear Safety

Safety is relative – there is no such thing as absolute safety. But people's perception of the relative hazard of different human activities is far from rational. A simple example serves to illustrate this point. Most people are aware of statistics which show that the risk of death or serious injury while travelling on the roads is far greater than while travelling in a modern aeroplane. But most of us experience a sharp increase in pulse rate and probably sweaty palms when we are in an aeroplane as the engines wind up at the start of the runway or as it makes a final approach to landing. On the other hand we will sit back and relax on the bus taking us to or from the airport – as long as we are not late.

Psychiatrists tell us that phobic fear is a very common human behaviour and is not related in any way to mental illness. It is involuntary and strong and is not concerned so much with what is happening now but rather with 'what if' worries. Even the most intelligent people have great difficulty in saying just what it is they are afraid of. It cannot be dismissed, but one of the best ways of dealing with it is to try to acquire a greater familiarity with the source of the fear and, if it is related to advanced technology, to attempt to get at least a superficial understanding of that technology.

The nature of radiation

Generally speaking, nuclear radiation is associated with changes to the internal structure in the nucleus of an atom – such as the complete nuclear fission or radioactive disintegration in which one or more of the particles in the nucleus split off spontaneously. In this respect it is not very different from light radiation and X-rays which are generally associated with changes in the orbits of the electrons flying around the nucleus of atoms, or for that matter heat radiation which is associated with the structural changes of atoms in molecules during the course of chemical reactions.

There are four primary types of nuclear radiation – alpha, beta, gamma and neutrons. Alpha radiation is composed of relatively large fragments consisting of two neutrons and two protons breaking off from the nucleus of the atom. Because of the size of the fragments and the fact that each alpha particle has a double electrical charge, this type of radiation does not travel very far before it comes to rest and becomes a neutral helium atom. The thickness of a pair of gloves would usually be sufficient to prevent the radiation reaching your hand. If, however, the radioactive element which is emitting the alpha particles is somehow ingested into the body, the fact that all the radiation would then be absorbed by the body means that the alpha radiation, with its double charge and denser ionization track, is relatively more damaging than more penetrating radiation with a less dense ionizing track. The main precaution against alpha-emitting radioactive substances is, therefore, secure containment to ensure that they cannot get into the body.

Beta radiation is composed of streams of light electrons. It is only distinguished from electrons which might be stripped from the orbits around atoms to create an electrical current, because the electrons are produced by transformation of particles in the nucleus of the atom and are emitted at relatively high speed. Because of the much smaller weight and charge of the electron, beta radiation is more penetrating than alpha radiation, but because each electron carries a negative electrical charge, not all that much more penetrating. Typically, a centimetre of a material such as perspex or aluminium would be sufficient to stop this radiation. Again, containment of beta-emitting radioactive elements to prevent them finding their way into the body is important, although they are somewhat less damaging than alpha emitters.

Gamma radiation is in effect one of the ways by which an atomic nucleus gets rid of surplus energy when its structure has been changed. It is therefore emitted at the same time as alpha or beta radiation or nuclear fission. It takes the form of electromagnetic waves similar to radio waves, microwaves or X-rays, and is also similar to heat and light radiation. But of all these different types of electromagnetic radiation, gamma radiation tends to be the most penetrating. It has no associated mass or electrical charge and can require tens of centimetres of lead or a metre or so of concrete to reduce its intensity to insignificant levels.

Neutrons, as has already been described, are emitted when the nucleus of a heavy atom, such as uranium, is fissioned but they are also produced occasionally by less dramatic radioactive transformations of elements.

When they are emitted from the nucleus, neutrons emerge at high speed – and are said to be fast neutrons. Since they have no electrical charge these fast neutrons are very penetrating and it might take several metres of concrete around a nuclear reactor to reduce their intensity to insignificant levels. On the other hand, once slowed down – or moderated – neutrons are fairly easily absorbed.

Another feature of neutron radiation which distinguishes it from the other three types of radiation, is the fact that when neutrons are absorbed in different materials they can make the atom in which they are captured unstable and these atoms are then transmuted into a different element. Frequently the new atomic element is radioactive and will emit its own characteristic mix of alpha, beta and gamma radiation. This process is known as *neutron activation* and it is important to realize that neutron radiation is the only type of radiation which can make other materials radioactive in this way. Protection from neutrons is therefore needed to prevent them from creating radioactive substances *in situ* in the body.

It is not difficult to see that with these different characteristics of nuclear radiation, the specification of meaningful numerical quantities is a complicated business. Basically, however, one is concerned with two quantities – the amount of radioactivity and a measure of the amount of damage that the emitted radiation will cause in human tissue. The unit used to specify quantity of radioactive material is the Bequerel (Bq). It is a measure of the number of atoms of a given radioactive species which are likely to disintegrate every second, irrespective of the mix of radiation emitted when these atoms disintegrate (previously the unit of measurement was the Curie). The ability of radiation to ionize matter is its main source of damage. This can be measured in units known as Grays (Gy) but for the purposes of measuring the dose of radiation to man the usual unit is the Sievert (Sv) because this is a composite unit which takes into account the absorption of the different kinds of radiation in human tissue and the resulting damage per unit of radiation (previously the units used were the Rad and Rem, each one hundredth of the new units).

A very detailed knowledge of the radiation characteristics of different radioactive substances, as well as their bio-chemical characteristics, is needed in order to specify acceptably small Becquerel quantities for ingestion but if one is considering an external source of radiation it is possible to specify doses directly in Sieverts, regardless of the source of radiation.

There is reliable evidence from the Japanese atom bomb survivors to indicate that if 100 people were to be exposed to a radiation dose of around 4 Sv then half of them would be expected to die from radiation sickness within a week or two and another three or four of the survivors would be likely to contract cancer later in life as a result of the exposure to radiation. Below 1 Sv the risk of early death would be very small but some burns and radiation sickness would be likely. There is biological evidence to suggest that the risk of later cancers would decrease in proportion to the dose – that is, there would be a linear dose-effect relationship – down to a level of about 10 mSv (a mill-Sievert is equal to one thousandth of a Sievert). This means that if a million people were somehow to be exposed to 10 mSv, about 150 of them would be expected to contract cancer over and above the natural rate of cancer in such a population, which is around 165 000. Below this level of radiation there is some question of whether the linear dose-effect relationship continues down to very low levels. There may well be a threshold level below which the body is capable of repairing any damage caused by radiation. But it would take massive experiments involving tests with tens of millions of animals to prove the point one way or the other. So it is pessimistically assumed that the linear relationship continues down to very low levels (Figure 4.1) and that any level of radiation dose has a probability of causing some additional cancers if you expose a large enough population.

Figure 4.1 Relationship between radiation dose and cancer risk with different possible extrapolations at low dose levels

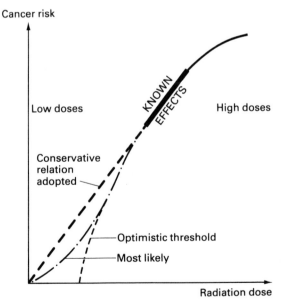

We are of course all exposed to a background radiation coming from outer space (cosmic radiation) and from natural radioactive substances in the ground and building materials around us and in our food (Figure 4.2). And we are exposed to radiation in the course of medical treatment with X-rays. Most of us are exposed to about 1.8 mSv in the course of a year due to background radiation and a further 0.5 mSv from medical X-rays. On the basis of the linear dose relationship this radiation would be expected to give rise to about seven additional cancer cases per year in a population of one million. It should, however, be mentioned that the level of natural background radiation varies from place to place around the world by a factor of two or three times. There is no evidence of increased incidence of cancer in the areas of higher background radiation, which tends to support the view that the assumption of a linear dose relationship at low levels is, indeed, very conservative.

Radiological protection

The dose limit recommended by the International Commission on Radiological Protection (I.C.R.P.) for workers in the nuclear industry is 50 mSv in any one year and for the general public the recommended level is 5 mSv. But there is an additional recommendation that dose levels should be kept as low as can be readily achievable below these limits. In

Figure 4.2 Proportion of radiation dose from natural and man-made sources for an inhabitant of a typical industrial country

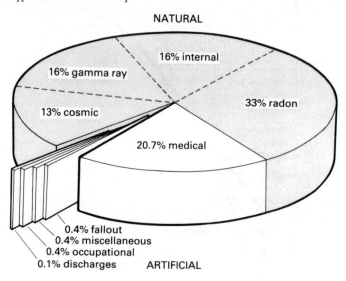

practice the average levels of exposure to nuclear plant workers are kept below 5 mSv per year.

In addition to specifying these general limits on radiation exposure for workers in the nuclear industry and the general public, the I.C.R.P. and radiological protection bodies which are responsible for setting legal limits in different countries, have established a list of detailed limits for different kinds of radioactive substances and for the exposure of different organs of the body. A major new profession, called health physics, has been created to administer these limits and at any establishment where radiation is encountered, whether it be a research laboratory, a hospital or a nuclear power station, there will be a team of health physicists to keep an independent check on day-to-day exposure.

It is important to realize that the dose limits recommended by I.C.R.P. and imposed by national regulatory bodies are not the levels above which there is a high risk of contracting cancer, still less a certainty of death. Rather they are the levels at which the conservatively estimated statistical probability of contracting cancer becomes so small that it is insignificant compared with the probability of contracting cancer from natural causes.

An example which illustrates this point is the case of twenty-six workers handling plutonium in the 1940s during the wartime development of the atomic bomb, under conditions which today would be considered very crude. They were exposed to environmental levels up to 350 times the presently permitted levels and ingested up to six times the permitted levels of plutonium. Admittedly this is a small sample of people but, more than thirty years later, the number of deaths among the group is about half what would be expected. Only two have in fact died – one being knocked down by a car and the other from a heart attack – and regular medical checks have revealed no evidence of harm from plutonium in the bodies of the survivors.

More generally, it is possible to say that if the permitted levels were significantly higher than they should be, we would today be observing a sharp increase in incidence of cancer among the thousands of people who have worked in nuclear research establishments around the world. Apart from the results of two rather questionable studies, the mass of information suggests that this is certainly not the case and the overall health record of nuclear workers is generally better than the rest of the population due, probably, to the very clean working condition and restrictions on smoking in nuclear establishments.

Protection against radiation can be provided in many ways. The first, and simplest, is distance from the sources of radiation. For example, if you were to stand within a metre of an unshielded source of radiation which was large enough to give you a particular dose in about one minute, then moving out to a distance of 8 m would increase the time to receive the same dose to more than an hour and at 40 m it would take over a day.

The second important method of protection is to provide some shielding between you and the source of radiation. This can take the form of protective clothing and perspex-fronted glove boxes for the handling of small radioactive sources, large walls of lead or concrete with massive windows and remote manipulators for dealing with more intense sources, and large thicknesses of water, steel and concrete around the core of a nuclear reactor.

Protection against the ingestion of any nasty substance can be provided either by concentration and encapsulation or by dilution and dispersion. In the case of radioactive substances, concentration on high grade filters or by evaporation of liquids and incineration of solids, is preferred because the volumes of material are small in the first place. But in the process of concentration there are always likely to be some residual traces of material in the stream of gas or water left over and in this case carefully controlled dilution and dispersion can reduce the concentration of radioactive materials to acceptably low values.

One valuable factor in the protection against the harmful effects of radiation is the ease with which it can be detected. A wide range of instruments has been developed to allow instantaneous measurement of levels of radiation far below the tolerable limits – in some cases it is even possible to detect the disintegration of a single radioactive atom. It is also possible to keep a check on the amount of radiation to which a worker has been exposed over a period by a simple device known as a film badge in which the darkening of a small piece of photographic film provides a measure of the total amount of radiation that has passed through it.

Nuclear safety measures

Development of nuclear energy for civil purposes dates from the early 1950s and the first Atoms for Peace conference in Geneva in 1955. By this time there was already a very full awareness of the potential dangers of radiation and radioactive materials. From the outset, therefore, the nuclear industry has taken very special precautions to protect against these dangers and safety has dominated all development work.

In a nuclear reactor the first priority if anything goes wrong is to stop the fission chain reaction in the core. Fail-safe systems, such as control rods, are usually held out of the core by electromagnetic devices so that if the power supplies are cut off the rods will drop into the core under the influence of gravity and shut down the reactor. Alternative back-up systems are provided on most reactors, especially in regions of the world subject to occasional earthquakes which might just cause a control rod to jam rather than fall into the core. These include hoppers which will release large quantities of small pellets of neutron-absorbing material into the core region, or systems which inject a neutron-absorbing liquid. Yet another back-up method used in some heavy water moderated reactors is a system in which the heavy water is held in the core region by a large pocket of gas in a dump tank below the reactor. Simply by releasing the gas pressure the moderator can be drained rapidly from the core region to stop the chain reaction.

In addition to these control systems, most reactors are designed in such a way that the chain reaction tends to cut itself off if the temperature rises above the normal operating levels – the self-regulation caused by a negative temperature coefficient which has been mentioned in Chapter 2.

Another important principle widely adopted in nuclear plants is redundancy of safety systems. This means, for example, that there will be several groups of control rods, each actuated by separate circuits, which are physically separated in the reactor building. Any one of the groups of control rods is designed to be sufficient to shut down the reactor.

Multiple redundancy is also used to improve the reliability of the control systems and prevent large numbers of spurious shut-downs which might otherwise result from false instrument readings. How this works is that each set of instruments measuring the critical conditions of the plant are replicated three, or even five, times and only if two out of three, or three out of five, alarm signals are received does the system automatically drop the control rods.

If a reactor has been operating for some time there will be some residual heat generation in the fuel elements even after the fission chain reaction has been shut down. This is due to the heat generated by radioactive decay of the fission products retained inside the fuel elements. Typically the heat generation in a reactor core would fall rapidly to less than 10 per cent of the operating power level in the first second after shut-down but would then fall off more slowly to 5 per cent in a minute, 1 to 2 per cent in an hour and to about 0.5 per cent after the first day. The second priority for the safety of the reactor is, therefore, to maintain sufficient cooling of the core to ensure that this residual – or 'decay' – heat does not cause the temperature in the core to rise to a level at which the fuel cladding might be damaged or melted.

In all types of reactors special features are provided to maintain some coolant flow through the reactor core after shut-down. The circulating pumps are usually equipped with 'pony' motors that will keep them going if the main drive motor fails; they may also have a flywheel to maintain rotation if there is a short interruption of power supply to the motors; and the power station will be equipped with a number of standby diesel generators which will cut in automatically if all external power supplies are lost.

Large power reactors are also provided with a number of alternative coolant systems which can be brought into action if the main flow is completely lost or if there is a major break in the primary coolant circuit, causing a loss of coolant fluid. These emergency core cooling systems (E.C.C.S.) differ on the various types of reactor because the rate at which the temperature in a reactor core will rise if there is a loss, or reduction, in the coolant flow will depend on the compactness (or power density) of the core, the size (or thermal capacity) of the moderator and the extent to which natural circulation of the coolant is effective in removing heat.

In a graphite-moderated reactor there is a large capacity for absorbing heat in the huge mass of the graphite blocks and even with natural circulation of coolant gas at atmospheric pressure it could take many hours for the temperature to rise to the point where there is a risk of damage to the fuel cladding. There is, therefore, a reasonable amount of time to start up emergency cooling systems or restore operation of the primary cooling system.

With sodium-cooled plants, such as the fast reactor, the power density of the core is very high and there is no mass of moderator to absorb heat. But on the other hand there is a sizable volume of the liquid metal coolant and natural circulation keeps it flowing through the core to remove the heat from the fuel. The main emergency cooling requirement is, therefore, to establish a secondary heat removal system to extract the heat from the primary coolant and this can also rely on natural circulation rather than circulation pumps.

In the case of water-cooled reactors the power density is intermediate between the graphite-moderated and fast reactors, and the ability to establish natural circulation could be complicated if there is also a loss of pressurization of the coolant and resultant boiling of the water. The emergency core cooling systems therefore have to be more elaborate and fast acting. Two systems are provided: the first uses high pressure gas in accumulators to inject cold water into the core region if the coolant system is still at a high pressure; the second uses robust pumps which take over at lower pressure and inject cold water first from large standby tanks and subsequently from a sump below the reactor where any water leaking out of the system would collect. Two to four redundant lines of both high pressure and low pressure injection systems are provided with any one of them being sufficient to keep the reactor core in a safe state.

Nuclear Safety

As well as ensuring that the chain reaction can be shut down and adequate core cooling maintained, a major safety consideration in the design and operation of nuclear power plant is containment. In practice this is a multitude of physical barriers, clean-up systems and operating procedures designed to ensure that if anything does go wrong, radioactive materials will be contained or their release greatly reduced by each of the successive barriers. And closely related to these protective measures is a vast network of radiation monitors inside and outside the plant to give early warning of any radioactivity where it should not be.

The main barriers have been mentioned in an earlier chapter. They include: the fuel material itself which, especially if it is a ceramic oxide, tends to retain most of the radioactive fission products; the fuel cladding made from the highest quality tubing and extensively tested by a variety of advanced testing techniques; the coolant, especially if it is water in which most radioactive elements dissolve; the reactor vessel and coolant circuits which are designed with a large margin of safety and manufactured and tested to the most stringent levels; the concrete vaults around the reactor and its coolant circuits which as well as providing direct protection against the radiation produced by the fission reaction in an operating reactor, provide a partial barrier against escaping radioactive materials and large cool areas of wall on which many vapourized substances would condense; and the building around the plant which, in the case of a water-cooled reactor, is usually a massive double-walled containment building capable of withstanding the maximum likely excess pressure from escaping steam inside and the impact of crashing aircraft on the outside (Figure 4.3).

It is one thing to provide all these safety features but another to ensure that they are all adequate and performing satisfactorily. This involves the most searching quality control at all stages of design and manufacture of every item of plant and equipment – in particular the fuel and the primary coolant circuits. When the plants are operating, carefully planned programmes of maintenance and inspection are undertaken as a regular routine.

Then there is a vast complex of control and monitoring instruments to allow the status of every item of equipment to be displayed in a central control room and to provide warnings of any malfunction. In some of the earlier nuclear power plants, this mass of control and instrumentation presented something of a problem in itself because there was so much of it that interpretation of alarms by the operators required an exceptional understanding of the whole plant. Fortunately, the development of modern computer technology has provided an answer to this problem in the form of data analysis systems which can rapidly analyse fault conditions and indicate to the operator appropriate corrective action. In fail-safe situ-

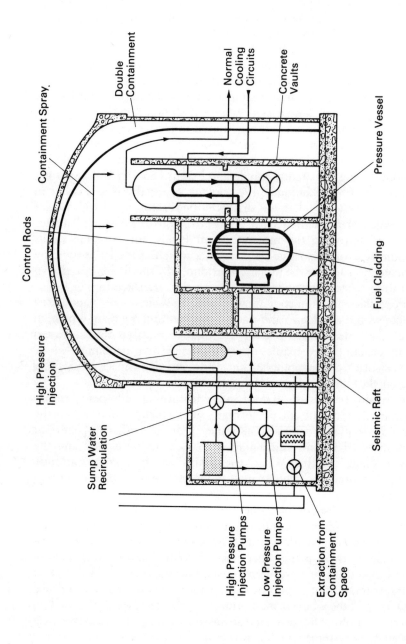

Figure 4.3 Some of the protective barriers and safety features of a typical pressurized water reactor

ations, appropriate action can, of course, be taken automatically, but there is an increasing trend to allow automatic systems to take other prompt actions to reduce the plant to the safest possible condition.

Careful consideration is given in the design and layout of plants to ensure that they are capable of withstanding the most severe seismic shocks likely to be encountered in the region where they are located. The normal procedure prior to construction of a plant is to study in detail the seismic record of the area, even if it is a region not normally associated with earthquakes, and then to design every component in the plant so that it will continue to function safely if it is subjected to a shock several times worse than anything previously encountered in the area. It is even possible to design reactors with sufficient safety margins to operate in areas of the world where earthquakes are relatively common. Apart from the provision of special shut-down systems, already mentioned, additional measures will include shock-absorbing mountings for all important items of plant and, in extreme cases, the building of the whole plant on a platform of rubber designed to absorb seismic shocks.

Clean-up systems are important in all nuclear plants, both during normal operation and to deal with accidental leaks and spills. The most important of these are comprehensive ventilation systems for the containment buildings, which are designed in such a way that the air pressure inside is less than it is outside and any small leakage will therefore tend to be inwards rather than outwards. The air extracted from the building by the ventilation system is then passed through very high grade filters before being discharged through a tall stack. Sensitive radiation monitors are provided to give an early indication of any build-up of radioactive material on the filters and to give an alarm if any radioactivity gets through the filters and up the stack.

The primary coolant circuit of a nuclear reactor will also be equipped with a variety of clean-up systems designed to remove the relatively small amounts of radioactive material which are produced by neutron activation of impurities in the coolant. A portion of the coolant flow, or in some cases the whole coolant flow, is passed continuously through filters and, in the case of water coolant, through columns of ion exchange resin which provide a powerful method for removal of all traces of impurity from the water. Special clean-up systems have also been developed for the liquid metal sodium coolant in fast reactors.

In the course of these routine clean-up operations, small amounts of radioactive material in gaseous form are likely to be extracted from the coolant or produced in the ventilation air near the reactor. These are not removed by normal filters and account for the only routine discharge of radioactive material to the environment. The amounts discharged are, of course, very carefully controlled and monitored to ensure that they are well below authorized limits. The gases are also inert in nature – this is why they cannot be easily removed by filtration – and are therefore not absorbed in the bodies of people or animals who may breathe in minute quantities. Routine discharges of this kind from nuclear plants are limited in such a way that they could not contribute a dose of more than 0.01 mSv per year – or less than 1 per cent of the dose from natural background radiation – to the population living around a plant, and in practice the requirement to keep levels as low as is readily achievable below authorized limits, ensures that actual emissions are very much smaller.

Regulation

Nobody would expect ordinary members of the public or, for that matter, individual experts, to judge whether a large and very complicated nuclear plant is safe. We rely on regulatory bodies who exercise a powerful independent control over the design, construction and operation of the plants just as we rely on airworthiness authorities for detailed evaluation and control of civil aircraft. The regulatory organizations which have been set up in different countries vary considerably in detail according to the governmental and legal structure of the individual countries, but they tend to follow roughly comparable steps before issuing a licence for a plant to operate.

The first stage, undertaken by the electric utility or commercial organization applying for a plant licence and assisted by the main contractors, is the preparation of an extremely detailed preliminary safety report. Either included in the safety report or as a separate report, there will be detailed consideration of the impact upon the local environment of the plant during normal and abnormal operations. Review of the preliminary safety reports by the regulatory bodies is likely to take anything from one-and-a-half to five years and will involve many hundreds of man-years of detailed analysis, both by the regulatory bodies and by the applicant in response to questions from the regulatory bodies. Many other review and advisory bodies are also likely to be consulted in this process, including special committees of independent experts and the well-established engineering standards bodies.

Another important stage in this preliminary review process is some form of consultation with the local population but here the exact procedures vary considerably from country to country. In the United States and Britain, for example, they may extend to a public inquiry with full adversary procedures, in which applicants and objectors can indulge in legal-style cross-examination. In France and Germany, on the other hand, public intervention at this stage is limited to a right to inspect all the documents associated with the planned project and to lodge written questions which must by law be considered and answered.

When a project gets through this rigorous preliminary review process a construction permit will be issued, usually by a central government department on the advice of the regulatory bodies, but in the case of Germany, first by a federal and finally by a state government department. In practice the construction permits are often issued in stages to provide a legal means for the regulatory bodies to halt work if for any reason they are dissatisfied with how it is being carried out and also to allow a certain amount of review and component testing work on later sections of the plant to be carried out in parallel with the early construction work. The regulatory bodies also specify very demanding standards of quality control which must be met during construction both at the plant site and in the many factories supplying equipment for the plant, and well-established inspection agencies are used to ensure that these standards are met.

Before an operating licence is granted for a nuclear plant a final detailed safety report has to be prepared and carefully reviewed by the regulatory bodies. This process of submitting and reviewing the final safety report can start anything from six months to two years before the scheduled operation of the plant. The operating licence is also likely to be issued in two or more stages – the first stage permitting the loading of fuel into the reactor for lower power tests and subsequent stages allowing operation at progressively higher power, depending on satisfactory results from the lower power tests.

Once a plant has a licence for full power commercial operation, the regulatory body is still responsible for ensuring compliance with the strict safety standards laid down in the licence. The operator is legally obliged to report any abnormalities in operation and to permit regular inspections by the regulatory body.

All countries operating nuclear power plants follow licensing procedures along these general lines and, even in cases where plants of well-established design have been imported, the national regulatory bodies have insisted upon making their own independent review of the safety of the project before issuing licences. In practice, however, the detailed procedures differ markedly from country to country – nowhere is it easy to get nuclear plant licences but in some countries the regulatory requirements and administrative procedures associated with the review process have escalated to a bewildering level. It does not follow, of course, that by making the licensing procedures more difficult you will achieve a safer plant, but there seems to be a belief – probably mistaken – in some countries that opponents of nuclear power will be reassured if the regulatory bodies are seen to be making life as difficult as possible for prospective plant operators.

There are also some fundamental differences in approach according to the circumstances in different countries. In the United States, Germany and Japan, for example, where there are a large number of private electricity utilities building and operating power stations, the regulatory bodies have established a very exhaustive and detailed set of criteria with which the plant designer and operator have to comply. In Britain and France, on the other hand, where there are large nationalized electricity utilities, the regulatory bodies have set general guidelines for overall safety standards and the utilities have to present a detailed case in their safety reports to show how they will meet and, if possible, better the general guidelines. Both approaches are capable of ensuring a very high standard of safety. There has been some criticism of the detailed criteria approach because it can encourage a 'checklist' approach among designers where, as soon as they have complied with all the difficult requirements, they assume that the plant is safe enough. The alternative approach puts greater onus on the designer to establish that his plant is safe enough but the subsequent review process relies to a greater extent on *ad hoc* discussions with the regulatory body and these discussions are less visible to the public.

How safe is safe enough?

Designing safety systems and assessing and testing their adequacy for something as complex as a nuclear power station is a major preoccupation of a large number of people in the nuclear industry and in the independent regulatory bodies. From the outset the designers have attempted to think of everything that could possibly go wrong and have then provided protection systems and back-up systems to prevent this happening and to minimize the consequences if it does. Likewise the independent assessors search for anything that the designer might have overlooked. But both admit to human fallibility. So a further approach that is widely used is to postulate a worst conceivable accident condition, even if one cannot see how such an accident could come about. This is known as the *design basis accident* and is best illustrated with an example.

The primary coolant circuit of a reactor is designed with a large margin of safety so that it should not fail, but the question is asked: if something has been overlooked and it does fail what is the most severe

Nuclear Safety

form of failure? The answer is an instantaneous guillotine burst of a pipe at its most critical point and displacement of the two broken ends so that a full flow of coolant can come out of each. Nobody can say how such an accident could actually happen but it is not possible to conceive of a worse situation so this is adopted as the design basis accident for the coolant circuit. The containment building around the reactor is then designed with sufficient strength to withstand the pressure from such an extreme break in the coolant circuit and the emergency core cooling systems are designed with sufficient capacity and a fast enough response to ensure adequate cooling in spite of the huge loss of coolant.

This particular example of a design basis accident is one of the most formidable that the designer has to face, but a similar approach is used at all levels throughout a nuclear power station. It certainly ensures that very adequate provision is made for the worst conceivable situation and is the basis for saying that the most likely consequence of a serious accident at a nuclear power plant is that nobody will be injured, even though the plant itself may be very severely damaged.

One drawback of the design basis accident approach in setting safety standards is that it focuses a lot of attention and a huge amount of design effort on very large plant failures and, although it is barely possible to conceive how they could happen, the safety experts who are involved in detailed consideration of the consequences tend to talk about them as if they are everyday occurrences. The familiar jargon of extreme accident scenarios is taken up readily by opponents at public hearings and meetings, and reported dramatically by the media. This contributes to fear and apprehension among the public rather than stimulating confidence in the fact that the worst imaginable has been taken care of.

A more recent method of assessing the adequacy of nuclear safety makes use of a technique called *probability analysis* – this is also a technique which has been used quite widely in the aircraft and aerospace industry. An attempt is made to assign a statistical probability – or chance – of failure to every item of equipment within a plant, and sequences of possible failures are built up with their associated probabilities into what are known as *fault trees*. If the sequence of failures represented by a fault tree ends up in serious accident consequences then it is possible to work out the probability of those consequences and also to identify the most significant failures in the chain of events. Applying probability analysis to a complete plant enables one to build up a table of probabilities for different classes of accidents, such as a small loss of coolant, a large loss of coolant, a core melt-down or a major release of radioactivity.

It is possible to extend the analysis, for those classes of accidents that would result in a release of radioactivity, and by considering the statistical probability of adverse weather conditions and the distribution of the local population it is possible to arrive at the risk of serious injury or damage to property. These risks can then be compared with other natural and man-made risks. An example of such a comparison from one of the first comprehensive probability studies of nuclear power plants in the United States is shown in Figure 4.4. It is this type of comparison that

Figure 4.4 Probabilities of different types of accidents causing large numbers of fatalities

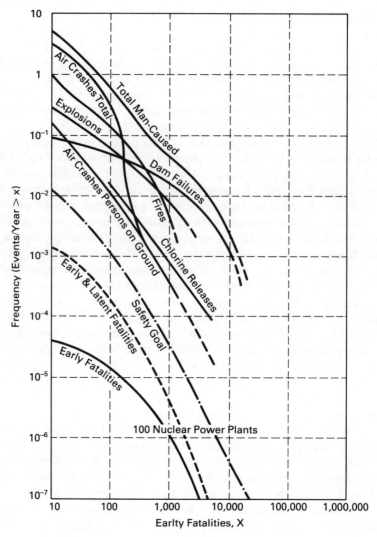

supports the view that, not only are nuclear power plants less likely to cause small everyday injuries, but they are less likely than other human activities to cause worst imaginable types of accidents.

But apart from helping to establish the relative safety of nuclear plants, probability analysis is a very useful tool for the designer. A comprehensive analysis of a complex plant will reveal those components and items of equipment whose probability of failure is most significant in contributing to a chain of events which could end up with a serious accident. The designer can then improve or duplicate that section of the plant and know that it will have the greatest effect in reducing the probability of the serious accident.

Probability analysis can also help to improve the reliability of operation of a plant by indicating those failures which have relatively small consequences but rather high probabilities of happening. In other words the designer is able to get a more balanced view of the whole plant than he does when he just looks at a design basis accident.

The difficulty with probability analysis is obtaining the initial statistical information about failure rates of different items of equipment such as pumps, pipes, and pressure vessels. Although a good deal of information is available from many years of experience in general engineering it is not always directly applicable to nuclear applications. In practice it is often on the pessimistic side because nuclear plant and equipment has from the outset been manufactured to higher standards of quality control than is found in traditional engineering. The amount of information about failure rates of individual items of equipment is rapidly growing as a result of comprehensive reliability reporting from operators of nuclear power plants in most countries.

At the moment, however, regulatory bodies do not have sufficient confidence in the accuracy of probability analysis to use it for setting the safety standards. Rather, they are tending to use a combination of the design basis accident approach for clearly identifiable types of failure and design target probabilities as an overall guideline for parts or all of the plant. Such a guideline might, for example, be aimed at producing a design such that the total predicted frequency of accidents which could give rise to uncontrolled releases of radioactivity to the environment, resulting from some or all of the protective systems being breached or failed, should be less than 10^{-6} (one in a million) per reactor-year and that a single accident leading to an uncontrolled release should be less than 10^{-7} (one in ten million) per reactor-year. A few examples of common everyday risk which most of us accept without too much concern are shown in Figure 4.5 and provide an indication of what a tough target is set for nuclear safety.

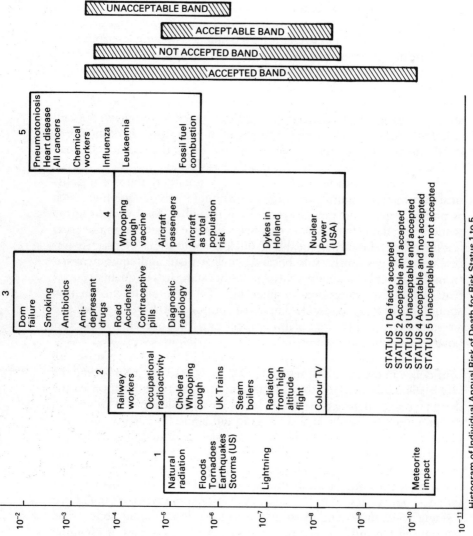

Figure 4.5 Classification of different risks according to public acceptance and actual acceptability

1 Centrifuge enrichment at a British–Dutch–German plant of Urenco

2 PWR fuel assemblies at the plant of Franco–Belge de Fabrication Combustibles

3 Sample of borosilicate glass of the type used for fixing high level radioactive wastes

Accidents do happen

Contrary to occasional media reports, the nuclear industry has never claimed that accidents will not happen. On the contrary, the basis of safety philosophy is that unexpected failures of plant and equipment will occur and human operators will make silly mistakes. The objective has been to ensure that the consequences of the resulting accidents will be minimized. By their very nature, any accidents involve a number of surprising and unexpected contributory factors and in most industrial accidents important lessons have been learnt from past accidents. But if loss of life or serious injuries are involved this is a very expensive way to learn lessons. Perhaps the greatest achievement of the civil nuclear industry over the past twenty-five to thirty years is that these safety lessons have been learnt without any loss of life or serious injuries directly attributable to nuclear radiation. But there have been plenty of minor incidents and a small number of accidents which have involved serious damage to nuclear plants. These incidents and accidents have received public attention which is usually out of all proportion to the seriousness of the consequences and this has become almost as great an incentive for the industry to learn the appropriate lessons as loss of life has been in other spheres of activity.

There have only been two nuclear accidents which have resulted in significant release of radioactivity beyond the immediate vicinity of the plant. The first of these was at Windscale in Britain in 1957 and the more recent accident was at the Three Mile Island (T.M.I.) nuclear power station in the United States. The Windscale accident involved an early air-cooled reactor used for military production of plutonium and, although it had some similarities to the later commercial gas-cooled power reactors, it would certainly not have met the safety standards now set for the civil nuclear industry. The air coolant was simply blown through the reactor core and out into the environment through filters at the top of tall stacks, rather than being pumped around a closed circuit. During an annealing operation, the fuel cladding melted and permitted combustion of the uranium metal fuel elements. In spite of this situation in which air was being blown through a burning core and discharged through filters, the release of radioactivity was confined to gaseous fission products and was not sufficient to cause any immediate harm. More recently a theoretical study of possible long-term effects on the British public has suggested as an upper limit of possible consequences of the Windscale accident that there may be, over a period of forty years, a total of about 260 additional cases of thyroid cancer of which 13 would be expected to be fatal. But there is no way that the actual occurrence of any such cases can be positively identified against the natural incidence of thyroid cancers which, over the same period, would total nearly 27 000 cases with over 1 300 fatalities.

For commercial gas-cooled reactors the important lessons learnt from this early accident were, first, a better understanding of the energy release from the graphite moderator and, secondly, the need for better instruments to tell the operators what was happening inside the core of the reactor.

The Three Mile Island accident involved a smaller release of radioactivity to the environment but was of wider significance to the civil nuclear industry because it was a large pressurized water reactor built to modern safety standards. A chain of relatively minor equipment failures led to a severe accident largely because the operators misinterpreted instrument readings and switched off the emergency core cooling which, quite correctly, had started to operate automatically due to a loss of coolant. As a result the core of the reactor was for a period not covered with water and suffered extensive damage, although no melting occurred. The containment building proved to be very effective in retaining most of the radioactive water which escaped but, ironically, it had not been completely isolated because the loss of coolant was not large enough to create the excess pressure which would have caused all valves to close automatically and the operators did not for a while recognize the necessity to shut the valves. There was therefore a pathway through some of the auxiliary plant which allowed a small fraction of the radioactivity from the damaged core to be discharged through a ventilation stack. The subsequent alarm about the possibility of an explosive hydrogen bubble in the reactor vessel, which resulted in a decision to evacuate some pregnant women and children from around the plant, was later shown to be a totally unfounded fear.

Apart from taking obvious measures to ensure that the initiating equipment failures would not recur, the more general lessons that have come from the T.M.I. accident are the need to improve operator training, and avoid ambiguities in instrumentation which might lead to similar misinterpretation of what is happening inside the reactor. Procedures, which had already been adopted in some other countries, to restrict operators from overriding automatic safety systems until they have had time to check the situation properly, have been implemented more widely.

The T.M.I. accident was also a striking demonstration of the dangers of concentrating too much attention on the very extreme design basis accidents when developing safety systems and procedures for a nuclear plant. The accident was, in fact, nowhere near the severity of a design basis accident but a small loss of coolant accident was allowed to escalate to the point where severe damage was done to the reactor core. As a result a great deal of research and analysis work has since been carried out to ensure that there is a better understanding of chains of events that might cause small failures to develop into serious accidents.

A disconcerting feature of the T.M.I. accident was that the same initial equipment failure and ambiguous instrument readings had been encountered previously on two other plants in the United States. The earlier incidents, which were halted before they became serious, were duly reported to the Nuclear Regulatory Commission for the benefit of other operators but the administrative system was such that the operators of T.M.I. did not pay due attention to the actions they should have taken to prevent the same thing happening at their plants. As a result, great efforts have since been made – both nationally and internationally – to streamline incident reporting by nuclear power plant operators and, more particularly, to improve the analysis of the implications of such incidents and the communication of the lessons to others.

An aspect of the T.M.I. accident which has aroused special concern in other countries was the incredible media reaction. The information services of the plant operators and the Nuclear Regulatory Commission proved to be totally inadequate in dealing with the hundreds of reporters who descended upon the plant, all wanting an inside story. Apart from illustrating the disproportionate attention devoted to any nuclear incidents, this type of reporting does have direct safety consequences. It can stimulate panic among the local population and cause them to flee the district when the best action would probably have been to stay indoors where a normal building would provide the most effective protection, even against a large initial cloud of radioactivity. Orderly evacuation might only be needed later in the extremely unlikely event of serious long-term contamination of the surrounding area.

Operational safety

While protection of the public at large is, quite rightly, a major concern in consideration of the safety of nuclear plants, the day-to-day safety of operating staff in all types of nuclear installation, from research laboratories to reprocessing plants, is also a very major concern. The major protection systems are, of course, as effective for them as they are for the general public, but the operational staff are also likely to be involved in a great many jobs which could involve some exposure to radiation or contact with radioactive materials. These can be routine maintenance jobs around the plant or larger operations such as the removal of spent fuel from a reactor and all the operations concerned with its transfer to storage ponds or shipment to reprocessing plants.

As previously noted the record of the nuclear industry has to date been exceptionally good. There have been minor incidents but at worst these have only involved exposure to radiation marginally over the conservative operating limits and the workers concerned have been moved for short periods to other operations not involving exposure to radiation. Over the years, and with expansion of the use of nuclear power, it is recognized that the total number of operational staff exposed to these occasional excess levels of radiation will increase to a significant population. As a result a continuing, and apparently quite successful effort, is being made to reduce operational exposure levels by paying greater attention at the design stage of a plant to all the maintenance and repair operations likely to be needed during the life of the plant.

5

Politics of Nuclear Energy

The civil use of nuclear power was launched in the 1950s with a large, and well meaning, political gesture on the part of President Eisenhower – the Atoms for Peace programme. Since this very positive beginning the nuclear industry has had to contend with a great deal of political involvement at all levels, ranging from the changing attitudes of world leaders and the vote-catching antics of national politicians, to the intricacies of local politics and the threat from extreme political groups that have jumped on the anti-nuclear bandwagon to further more devious political objectives.

In the early days the development of nuclear energy attracted enthusiastic political support as a highly prestigious field of advanced technology which offered great potential benefits to individual countries and to the world at large. But with the advent of environmentalism in the late 1960s the subject became one of the favourites for politicians who wanted to be seen to be showing concern. One of the main problems with which the nuclear industry has had to contend is the fact that political positions – so often dictated by electoral considerations – have tended to change at a faster rate than they can respond. The period from original conception to operation of a nuclear power station is often in excess of ten years while governments frequently change at intervals of five years or less.

Non proliferation

Without question the most important political issue associated with nuclear energy is the problem of ensuring that the technology is used only for peaceful purposes. This was recognized at the time of the launching of the Atoms for Peace programme and resulted in the setting up of a United Nations agency, the International Atomic Energy Agency (I.A.E.A.) based in Vienna. Established at the height of the 'Cold War', the I.A.E.A. was remarkable in gaining the active support of both East and West in the common aims of promoting the use of nuclear energy for the benefit of mankind and establishing safeguards to prevent the spread of nuclear weapons. The promotion of the use of nuclear energy was not limited to nuclear power but extended into a whole range of parallel uses in medicine, agriculture, food preservation and pest control where radioactive isotopes and radiation techniques have many valuable applications. In this way the I.A.E.A. has also been very successful in helping and gaining support from the developing world.

The policy adopted by the world leaders at the first Atoms for Peace conference in 1955 was to declassify all basic nuclear technology required for peaceful uses with the one exception of the details of uranium enrichment processes. This was possibly motivated by realization of the fact that most of the principles were well known by scientists throughout the world and it would only be a matter of time and money before more countries developed their own capability in nuclear technology. How much better to ensure from the start that development of peaceful applications would take place openly and with many opportunities for international collaboration. The exception was made in the case of enrichment techniques because, although the basic principles were also widely known, the cost of independent development was considered to be so prohibitively high that individual countries would not attempt it. This argument could not, however, be applied to separation of plutonium from spent fuel taken out of reactors, because the chemical processes involved are straightforward.

The countries which had made the huge investment to develop enrichment for military purposes – notably the United States but also the Soviet Union, and later Britain and France – undertook to make supplies of enriched uranium available for research reactor fuel and for power plants, subject to a system of safeguards controls administered by the I.A.E.A. The safeguards system developed by the I.A.E.A. aims to keep track of every scrap of nuclear material by a detailed system of accounting, by periodic and random inspections of nuclear installations and with some physical techniques such as tamper-proof surveillance instruments or bonded stores.

The I.A.E.A. has been remarkably successful in establishing safeguards agreements with over 100 countries, including all but a very small number of those that have any kind of activity in the field of peaceful uses of nuclear energy. Nobody can claim that the system of controls and inspections is completely foolproof and there is a continuing effort to strengthen and improve I.A.E.A. safeguards. Nor is it envisaged that safeguards could prevent a determined government from diverting some nuclear materials from civil uses to military developments. But there is a high probability that any such diversion would be detected, or suspected, and this would open up the way for intense political pressure to be brought to bear on the country concerned to discourage it from following its military objectives.

The other major political development at the international level came in 1968 with the signing of the Non Proliferation Treaty (N.P.T.) by the United States, the Soviet Union and Britain. Since then some 115 countries have become party to the N.P.T. Essentially the N.P.T. strikes an agreement between the nuclear weapon states and the non-weapon states under which the latter undertake not to develop weapons in return for access to nuclear technology and assurances of nuclear fuel supplies for peaceful uses from the weapon states. Parties to N.P.T. also undertake to implement a safeguards agreement with the I.A.E.A. if they have not already done so.

The fact that there has been no widespread development of nuclear weapons in different countries over the past twenty years, although most industrial countries and several of the larger developing countries clearly have the technical competence, must be judged as a significant achievement for the I.A.E.A. safeguards system and the N.P.T. But there is still some way to go before there is general peace of mind and mutual international trust.

Most countries which are not party to the N.P.T. and have no safeguards agreement with the I.A.E.A. have no significant nuclear activities to worry about. This leaves two weapon states – France and China – and four other countries with significant nuclear activities – India, Israel, Pakistan and South Africa – outside international controls. Of these, France is subject to the safeguards controls of the European Economic Community and operates in the same way as other weapon states that are party to N.P.T.

As well as clauses establishing the right of non-weapon states to develop nuclear energy for peaceful purposes and the requirements for safeguards inspection, the N.P.T. contains a further clause in which the three sponsoring weapon states undertake to work for a general reduction in weapons stockpiles. The fact that little or no progress has been made in this direction was the source of considerable friction at the last N.P.T. review conference in 1980. A group of non-aligned countries, in particular, fiercely attacked the United States and the Soviet Union for failing to make significant progress in the SALT disarmament negotiations and claimed that the weapon states were not keeping to their side of the bargain implicit in N.P.T.

On the other hand, following the 1974 explosion of a nuclear device by India – said to be for peaceful purposes – there has been recognition, especially in those countries with the technical capability to supply nuclear equipment and services, of the need to strengthen the safeguards conditions covering such supplies to other countries. In passing it should be noted that at the time that the Indians were developing a nuclear device there were much publicized programmes in the United States, and to a lesser extent in the Soviet Union, to develop the use of nuclear explosives for peaceful purposes such as stimulation of gas and oil recovery from tar sands or excavation of irrigation canals. The Indian interest in projects of this kind, with real economic potential, was also publicized at international conferences but, without doubt, the actual testing of an explosive device has shown that India has all the necessary skills to develop nuclear weapons. International concern has been increased by the unwillingness of the Indian government to submit all its nuclear installations to I.A.E.A. safeguards inspection – what is referred to as *full-scope safeguards*.

One result of the effort to establish more effective safeguards in the aftermath of the Indian explosion was a get-together of the main nuclear supplier countries in a series of meetings in London to establish conditions for nuclear exports which go beyond the I.A.E.A. safeguards. The group, now generally referred to as the London Suppliers' Club, produced a list of guidelines which have been generally adopted by the members of the Club in the conditions imposed upon their export sales of sensitive nuclear equipment and services. They also reinforced the role of the I.A.E.A. in exercising safeguards control over individual sales but did not go as far as insisting upon full-scope safeguards before agreeing to supply equipment to other countries.

Another major development came in 1977 when a nuclear policy statement of President Carter called into question, for the first time, the desirability of commercial reprocessing and fast reactors on the grounds that their widespread development could increase the risk of proliferation and the immediate need for introduction was not apparent. Accordingly he announced a policy of self-denial, for the time being, in his own country and in the following year the Congress adopted the U.S. Nuclear Non Proliferation Act which imposed conditions to prevent other countries from following this fuel cycle route with fuel or services of American origin except with prior consent from the U.S. government.

This action produced a storm of protest from other countries, particularly in Europe and Japan, where many nuclear power programmes were dependent on American enrichment services, because of the unilateral nature of the act and the fact that it revoked existing bilateral agreements and long-term fuel supply contracts to countries which were party to the N.P.T. and had never shown any inclination to divert material from civil nuclear power plants to military developments. Particular resentment was felt because the need to reprocess fuel and recycle plutonium in thermal or fast reactors could be more immediate in Europe and Japan. Unlike the United States, these regions have no significant domestic uranium resources. The eventual ability to recycle is seen as a means of using uranium much more efficiently and so avoiding the sort of supply problems that have been encountered with oil.

Largely in response to the worldwide criticism of the Carter policy, a massive international study exercise was mounted. Known as the International Nuclear Fuel Cycle Evaluation (I.N.F.C.E.) it lasted for two years and considered in great depth every aspect of different nuclear fuel cycles. The intention was to provide a better international concensus on the technical aspects of the fuel cycle alternatives and so form a stronger basis for political decisions of the kind that President Carter had attempted to implement. With over fifty countries, from both the East and the West, and a number of international agencies participating it is perhaps remarkable that I.N.F.C.E. was successful in producing a number of consensus reports containing a huge amount of technical detail on this very complicated subject.

A number of important conclusions were reached by I.N.F.C.E. To start with there was a strong endorsement of the necessity for civil nuclear power to meet the world's present and future energy requirements. It was recognized that all nuclear fuel cycles, including the once-through fuel cycle with no reprocessing, present some risk of proliferation. So-called 'technical fixes' which had been examined as a possible way to make sensitive areas of the fuel cycle more proliferation-resistant were rejected for the most part because they tend to introduce as many problems as they solve. But it was concluded that none of the established fuel cycles, including those involving reprocessing and fast reactors, presented insurmountable problems or proliferation risks that were significantly different from alternative fuel cycles.

A number of measures, involving many institutional arrangements rather than technical fixes, were suggested by I.N.F.C.E. as the best route for the development of tighter controls. Most important of these was endorsement of the I.A.E.A. as the best organization to operate the system of safeguards and the view was that the best way forward was to build on the existing arrangement which, as multinational agreements go, is working remarkably well. It was suggested that as far as possible it would be desirable for reprocessing to be undertakn in a small number of large plants in countries with major nuclear power programmes and there might also be some merit in establishing regional fuel cycle centres, incorporating large reprocessing plants, under multinational ownership and operation, to serve the needs of a number of smaller countries.

The international safeguards system of the I.A.E.A. suffered a severe blow in 1981 when Israel attacked and destroyed a research reactor in Iraq just before it was put into operation. This politically-motivated action was particularly regrettable because it was an attack by a state which is not party to N.P.T. and which does not submit its own facilities to I.A.E.A. safeguards inspection on a state that has signed the N.P.T. and has opened all its facilities, including a smaller research reactor which has been operating for many years, to inspection. Subsequent analysis suggested that the Iraqis had no intention of misusing their new reactor for military production of plutonium and that any misuse would, in any case, have been detected by the safeguards inspection that was being put into effect by the I.A.E.A. This case shows that even though there may be effective institutional arrangements for international control of the peaceful development of nuclear energy, these are not enough in themselves to overcome the political differences and mutual distrust which dominate some of the more troubled parts of the world.

Plutonium politics

It will already be apparent that plutonium has dominated the discussion of proliferation problems. Historically, this is because it is produced to a greater or lesser extent in all nuclear reactors and can be separated from the spent fuel by well-known chemical processes to produce a relatively pure material with sufficient density of easily fissionable atoms to make explosives. There is now a view, supported by the I.N.F.C.E. conclusions, that perhaps too much prominence has been given to the problem of plutonium. The practical problems of separation of plutonium from highly radioactive spent fuel are proving to be far more difficult than might be thought from reading a textbook on chemistry. On the other hand new techniques for the physical separation of uranium isotopes offer the prospect, at least, of easier enrichment processes which could in theory be used for the production of uranium of sufficient enrichment to make an explosive. Thus, while the specific problems of plutonium are still being taken seriously, some increased attention is being directed towards alternative routes to nuclear weapons proliferation.

In dealing with specific plutonium problems, the most sensitive area is certainly the control of the pure material once it has been separated from spent fuel by reprocessing. In parallel with I.N.F.C.E. the I.A.E.A. therefore took the initiative to look into the possibility of storing the stockpiles of separated plutonium under international control. The idea is that once separated the plutonium would be held in a high security-bonded store, probably located at the same site as the reprocessing plant in which it was separated, and the I.A.E.A. would keep account of all material going into the store and control its release for legitimate civil uses. This would allow the operators of nuclear power plants to gain early benefit from recycling of unused uranium – which makes up the bulk of spent fuel – while leaving their separated plutonium in the internationally controlled store until such time as they could establish a need to withdraw it for use in fast reactors or associated research and development work in their countries. The concept of international plutonium storage was endorsed by I.N.F.C.E. and countries with nuclear power programmes have since been trying to work out details in an international working group. The hope is that a satisfactory scheme could be introduced in the early 1980s which should be in time to deal with the start of large-scale commercial reprocessing of fuel from light water reactors in plants in France, Britain and possibly the United States.

The main hurdle to be overcome in establishing international institutional arrangements such as plutonium storage is mistrust of the major supplier countries that has been generated among smaller and non-aligned countries. The Carter Administration, and to a lesser extent the actions of the London Suppliers' Club, have shown that long-term international agreements for the supply of nuclear fuel cycle services, which could become crucial to a small country's energy needs, could be revoked unilaterally. Despite many assurances, non-aligned countries want to be sure that if plutonium separated from the spent fuel was to be deposited in an international store in a major supplier country the rules would not be changed when the time came for them to withdraw it for legitimate civil purposes. A parallel initiative by the I.A.E.A. to establish a Committee for Assurances of Supply to look into schemes such as a 'nuclear fuel bank' which would attempt to ensure that supplies of nuclear fuel would be made available to keep vital power plant going even in periods of political dispute, may help but it is also mistrusted by the smaller countries.

Before leaving the subject of plutonium, mention should be made of one particular misconception. This is the belief that introduction of fast breeder reactors will result in 'fast breeding' of the quantities of plutonium being produced in the world. It has led opponents of the fast reactor to coin the phrase 'plutonium economy' to suggest a situation in which there is a large international trade in plutonium to fuel fast reactors and, by implication, a risk that some of this plutonium will fall into undesirable hands. In fact the net production of plutonium associated with each unit of electricity production from a nuclear power plant, is somewhat higher for most of the currently used thermal reactors than it is for a fast reactor. This is because the bulk of plutonium created in the depleted uranium blanket around the core of a fast reactor would be recycled to fuel the core of the same reactor. Even under optimum conditions of breeding, only a small fraction of surplus plutonium would be produced which over a period of twelve to fifteen years could build up to a sufficient quantity to fuel one new fast reactor core. The relative balance of plutonium production in typical thermal and fast reactors is represented in Figure 5.1.

It is also important to remember that a fast reactor only breeds plutonium if depleted uranium is fed into the blanket around the reactor. If there was to be a world surplus of plutonium from the accumulated production of the large number of thermal reactors, it could be 'burnt' in fast reactors without breeding any surplus and in this sense a fast reactor could be considered as a very productive incinerator of plutonium. This course of action might not be very economical but it would be better than disposing of spent fuel, containing unused uranium and plutonium, as waste, after one pass through a thermal reactor.

Figure 5.1 Balance of net plutonium production in thermal and fast reactors

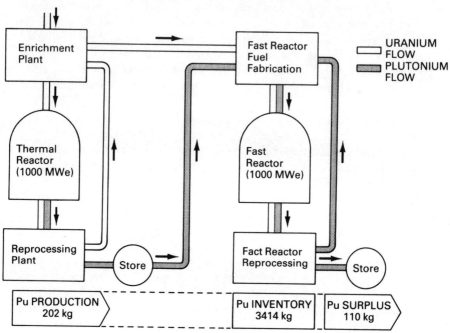

National politics

At first sight nuclear energy would seem to be an area of development that should receive enthusiastic support from most politicians at a national level. It offers greater national independence in energy supplies – even if uranium has to be imported the quantities involved and the diversity of sources of supply make it a good deal less politically sensitive than oil. Nuclear energy can also stimulate activity in a broad spectrum of advanced technology which will help the development of national industries. In the early days these attractions were indeed embraced by many political leaders in various countries and there were few if any dissenters, at least for the peaceful uses of nuclear energy. Later, however, a number of factors turned nuclear energy into a popular issue for political opportunism. Notable among these was the possibility of being seen to be responding to one of the favourite campaigns taken up by the burgeoning public interest groups. It is also seen as an area where politicians can try to respond to demands for more open government by exposing information which, it is claimed, originated under a 'cloak of secrecy'. Then there are the large sums of public money needed to support the development of civil nuclear power programmes, which is an obvious area for political argument. This rather cynical view of politicians' concern about nuclear issues is supported in part by the fact that it is usually only the minority political parties, often of widely differing political persuasions, which come out clearly on anti-nuclear platforms. Those nearer to the actual problems of governing an industrial country have, for the most part, come to the conclusion that they cannot manage without nuclear power though they often try to placate the noisy minority by adopting nuclear energy as the option of last resort.

In some countries politicians have also got involved in technical choices between alternative routes of nuclear power development. Strictly speaking they should not be called upon to make technical decisions but rather should seek clear advice, either from the experts of national research and development agencies, or from the nuclear and electric utility industries, or from independent bodies of advisory experts. The problem here has been that the advice from the different areas is frequently contradictory and often very complicated. Industrial factions supporting different types of reactor often make exaggerated claims for their own system and denigrate another when in reality both systems are probably very good but not perfect. Research and development organizations seeking funds to maintain a particular line of development are also likely to be over-optimistic about the time-scale and cost of commercialization. Although the nuclear industry and the research and development organizations tend to complain of undue political interference on technical questions they often have only themselves to blame for failing to present to government a clear technical concensus on which the political decisions, often involving the allocation of very substantial funds, can more easily be made. All too often, however, politicians have been called upon to arbitrate between conflicting bodies of expert opinion and they cannot really be blamed if their reaction as often as not is to put off the decision.

A third area where national politics has had a big impact on the development of nuclear energy is in the organization of industry. The scale of nuclear projects demands industrial organizations with large financial resources and a great diversity of technical skills. In many countries no one company could offer this and the early days of commercial nuclear power saw the formation of many industrial groupings. As the unit size of nuclear power stations increased, it also became apparent that the rate of ordering projects in most countries was insufficient to justify competing industrial groupings and all the duplication of industrial investment and technical resources that this would involve. At one time there were hopes that this problem might be resolved with one or two multinational groupings of companies competing in a large market such as Europe. But nationalism has prevailed and instead we have seen progressive merging of groups in individual countries with governments playing a major role in the painful reorganization process and in many cases having a direct involvement, in the form of substantial share holdings, in the industrial groups. The pattern of industrial development in different countries and the relative strengths and weaknesses will be dealt with in Chapter 8.

The political climate, of course, varies dramatically from country to country but a few examples will serve to show how nuclear issues have come to play an increasing role in national politics. Perhaps the most dramatic example is Sweden, where the nuclear power programme became a major issue in two general elections, caused the fall of one government and after a five-year period of intense argument was finally resolved by a national referendum. In 1975, the leader of the Centre Party, Thorbjorn Falldin, took a strong stand against the country's nuclear power programme in the general election. The Centre Party is one of three smaller parties forming an alliance to the right of the Social Democrats who, up to this time, had been in government for forty-four years. In the event the right wing parties gained a narrow majority in parliament and formed a government with Falldin as Prime Minister but with a coalition in which the other two parties supported the need for nuclear power. The inevitable conflict eventually split the coalition and

Falldin resigned at the end of 1978. He returned in a subsequent general election but only with an undertaking to his partners that he would abide by the conclusion of a referendum on the nuclear issue. The referendum was held in 1980 and approved a programme of up to twelve nuclear reactors for the country most of which are already operating or due to come on line during the 1980's. The issue has now largely disappeared from the Swedish political scene but it may re-emerge in the future since all the options in the referendum called for eventual phasing out of nuclear power at the end of the useful life of the twelve reactors. Decisions on whether or not to do this will have to be taken before 2010.

The problem that the nuclear industry has to face in dealing with changing political attitudes is illustrated by the example of Austria. When in government in the late 1960s the Christian Democrats sanctioned the building of the country's first nuclear power station. By the time it was ready to operate in 1977, a large public opposition movement had emerged and the Christian Democrats, now in opposition to the Social Democrat government of Bruno Kreisky, took up a position against operation of the plant. A subsequent referendum decided by a very narrow majority to reject operation of the plant and since that time it has stood idle in spite of the country's growing balance of payments problem caused by the cost of importing oil, gas and coal.

In Britain the need for a contribution to electricity production from nuclear power was never seriously questioned in the early days by the major political parties, in spite of the large coal resources and more recent discoveries of natural gas and oil below the North Sea. But on several occasions the question of technical choices between alternative types of reactors has become a heated issue at the political level, mainly due to failure of different factions within the industry to resolve their differences. The result is support from the Conservative Party for a change from gas-cooled reactors to pressurized water reactors – subject to the result of a public inquiry – and outright rejection of the pressurized water reactor by the other parties.

Local politics

Nuclear power plant projects have featured large in classical conflicts between central and regional governments especially in countries with a federal structure. In the United States the federal government, in spite of several challenges over the years from state governments, has maintained its preemptive powers in matters of nuclear power plant licensing and regulation. But this did not prevent state government of the largest and most prolific energy consuming state, California, from enforcing state regulations which in effect prevented the start of any new nuclear power plant construction projects for the best part of a decade. This in spite of the fact that a popular vote on an initiative to prevent operation of nuclear power plants in the state was defeated by a substantial two-thirds majority in 1974.

In the Federal Republic of Germany the conflict between the state (Lander) governments and the federal government on nuclear issues is particularly acute and very complicated. Under the German law the final responsibility for issue of nuclear licences rests with the state governments although detailed proposals have to be submitted to federal agencies for technical appraisal. At the federal level the main political parties have generally supported the need for nuclear power although with rather more enthusiasm in the coalition led by the Christian Democrats than that previously led by the Social Democrats. The party policy of the latter on energy was usually stated as 'priority for German coal, conservation, and limited development of nuclear power' but during its term of office it came into conflict on a number of occasions with state government led by the same coalition but opposed to the issue of nuclear licenses in their states. Under the coalition that took over the federal government in 1982, increased intrasigence can be expected from those state governments led by Social Democrats and in the states where the new Green Party – passionately opposed to nuclear power – has obtained the necessary 5 per cent of votes to gain parliamentary seats.

The situation in Germany has been further complicated by administrative courts at the local level imposing orders to halt construction on projects often on the grounds of technical issues on which they are not strictly qualified to pass judgement. Only by lengthy appeal to higher courts has the technical authority of the federal agencies been upheld and work allowed to continue.

Throughout the 1970s, successive Italian governments have attempted to implement a national energy plan which includes a sizable contribution from more nuclear power plants to deal with the country's increasingly critical shortage of electrical generating capacity. This should have been possible in spite of the frequent changes of government because the major political parties support the need for nuclear power in a country which has very few domestic energy resources. But regional governments have raised strong opposition to any specific plans to site nuclear power stations in their locality and only one new project of a planned ten or twelve was able to get under way during the decade. To overcome this problem the national nuclear agency has drawn up a map of prospective sites fairly evenly distributed throughout the country and the central government is having to negotiate special agreements –

usually involving substantial development subsidies – with regional governments to allow detailed site selection from two or more candidate sites in each region.

Political extremists

The growth in the number of political extremist groups and the increasing violence of their activities is a particularly disturbing feature of the world in general. Such groups are inclined to grasp at any major public issue as a possible means of furthering their devious political objectives. There is evidence of extremist elements infiltrating anti-nuclear movements in a number of countries, especially Germany, but it is only in Spain that a group has come out in open defiance of a nuclear power project. Even after the granting of partial autonomy to the Basque region, the militant faction of the Basque Separative Movement has continued to make attacks on a nuclear power station under construction near Bilbao. Terrorist attacks against the plant have resulted in the death of three workmen from an explosive, the death of a bystander in cross-fire with the police and finally the murder of two chief engineers.

Those concerned with the security of nuclear installation in other parts of the world have not been slow to realize that they are potential targets for extremist groups and have taken measures to increase protection. There is, however, no reason to suggest that nuclear installations are any more vulnerable than many other vital civil and military installations. In many respects the layers of protection provided to prevent the escape of radioactivity from a nuclear power plant are very effective in providing the plant with protection from outside attack. The radioactivity itself makes access to critical areas of the plant hazardous for anybody trying to place a bomb and the fail-safe nature of the control systems makes it virtually impossible for anybody to drive a reactor out of control except with very specialist knowledge of how the system works.

From time to time the opponents of nuclear power also present the spectre of terrorist groups getting hold of some highly enriched uranium or plutonium and constructing a crude home-made bomb. It is pointed out that the principles of a nuclear explosive are relatively simple and in an effort to prove this point science students have produced designs on paper which it is said would work. In practice, however, even if it was possible to obtain the uranium or plutonium in a sufficiently pure and concentrated form, a wide range of technical skills would be involved in converting it even into a crude explosive device and it is unlikely that anybody other than a team of highly qualified specialists would possess all the necessary skills.

There are, of course many other potential targets in the modern world which would allow extremist groups to cause just as much devastation as they would hope to cause with a home-made nuclear device, if it worked. It is questionable, however, whether large-scale devastation is an objective which would be in the devious interests of an extremist group. The pattern of terrorism around the world seems to suggest that they are most effective in creating terror among the public with a large number of attacks on different targets with relatively small conventional explosives which they are able to obtain all too easily.

6

Waste Management

Of all the issues surrounding the peaceful development of nuclear energy the question of waste management is the most persistent. While other issues come and go as the opponents of nuclear energy adopt the ploy of continually shifting their attack, there remains a much wider concern about the responsible management of radioactive waste materials.

Man has a pretty poor record of dealing with waste, whether it be urban sewage, general garbage or industrial waste, so it is not altogether surprising that the advent of a new industry whose waste products are endowed with the mysterious property of radioactivity should be subject to great suspicion. The apparent irresponsibility of man in dealing with problems in the past has usually been due to the very rapid growth of urban populations and the rapid development of new industries in which the waste production has grown to unmanageable proportions before the scale of the problem is appreciated. The problem is compounded by the attitude of mind which suggests that you should not have to spend a lot of money getting rid of waste.

In both these respects the nuclear industry is genuinely different. The scale of production of waste materials is manageable – to take an extreme hypothetical case, it would be physically possible to take the entire world production of dangerous radioactive waste and dispose of it in one disused mine. And, from its outset, the nuclear industry has devoted very considerable financial resources to research and development on all aspects of radioactive waste management.

Where the nuclear industry is subject to criticism is in taking a rather relaxed attitude on the need to detail final waste disposal plans – this it has done because the rate of accumulation of waste is manageable and there therefore appeared to be plenty of time to pursue further research and development before making a final choice of optimum disposal techniques. Due to the many complexities of waste management procedures the industry has also failed to satisfy the public that it has the situation in hand and is behaving responsibly.

Strenuous efforts are currently being made around the world to rectify these two shortcomings, but it should be recognized that, unlike urban waste or other industrial waste, it has not taken a major outbreak of disease or a pollution scandal to stimulate the intensified waste management activity.

Classification of radioactive wastes

According to media mythology all radioactive wastes are 'deadly'. In fact there is, of course, a complete spectrum of radioactive waste classifications ranging from low level liquid wastes, which are often less radioactive than widely sold brands of mineral water, to concentrated high level wastes which could cause severe burning and probable death if allowed to come into contact with people.

Radioactive wastes are generally categorized as low, intermediate or high level. These classifications are not precise and are used mainly to give an indication of the amount of shielding that is likely to be required when handling, transporting or storing the materials. Typically, however, high level liquid waste might contain more than 10 billion Becquerels per cc while low level liquid waste would contain less than 100 Becquerels per cc (a Becquerel is a very small quantity of radioactivity, representing the disintegration of only one atom per second).

Then there is the more obvious categorization into solids, liquids and gases which will have an important influence on the form of containment to be used and eventual disposal practices. A further distinction is made between beta-gamma wastes and alpha-containing wastes. If the waste contains only beta- and gamma-emitting radioactive isotopes the main concern is the provision of sufficient shielding to stop the bulk of the radiation, although isolation of chemically active substances also has to be considered. If there is a likelihood of some alpha-emitting radioactive isotopes being mixed in with the beta-gamma emitters then there is a need to take rather more stringent precautions to isolate the materials because, although it is very easy to stop alpha radiation in shielding, the alpha-emitting isotopes are significantly more hazardous if they escape and are inhaled or ingested. Fortunately most alpha-emitting isotopes are heavy atoms which do not normally occur in gaseous form, and secure containment of liquids and solids is relatively easy.

While these categories are widely used in discussing the overall provisions for waste management within the nuclear industry, it is in practice necessary to consider carefully the nature of each incidence of waste

Waste Management

material in deciding how to handle, store or dispose of it. A reference to 'only low level solid waste' may, for example, sound unduly dismissive but in practice anybody handling such material would have to satisfy themselves and regulatory bodies that they knew the precise nature and radioactive content, and had taken adequate precautions.

Bearing in mind the point already made about the very small physical volume of nuclear wastes compared with waste from other human activities, and the fact that all radioactive waste decays, the general approach to waste management is concentration and isolation rather than dispersion. Most high and intermediate level wastes therefore arise as a result of concentration of lower level wastes by techniques such as filtering, evaporation, incineration or chemical separation. On the other hand the majority of low level waste consists of the large volumes of slightly radioactive water or air left over after the bulk of radioactivity has been removed in the concentration processes. Subject to strict limits, this can then be disposed of by further dilution and dispersion into a large body of water or up a tall stack. To this should be added material which starts off as low level solid waste and, although it may be reduced in volume by compaction and incineration, the radioactive content is still sufficiently low to allow it to be disposed of in shallow land burial or deep sea bed sites where it will decay to totally harmless levels before there is any chance of dispersion.

We will deal here mainly with radioactive waste produced by operating nuclear power stations and the associated fuel cycle services. But it is important to remember that this is not the sole source. Radioactive waste, particularly low level solid waste, is also produced by hospitals, a wide variety of research laboratories, and industrial users of radioactive isotopes. In general, however, these other sources usually pass their radioactive waste to the major national nuclear research centres for appropriate treatment and disposal and it therefore becomes indistinguishable from the waste produced by the nuclear industry at large.

Table 6.1 gives the approximate amounts of waste associated with the operation of a typical 1000 MWe nuclear power station for one year, and to keep this in perspective Table 6.2 shows the quantities of waste associated with a year's operation of a similar sized coal-fired station.

Waste from fuel production

The first, and physically the largest, source of radioactive waste is associated with uranium mining operations – the mill tailings. The concentration of uranium in commercially mined ores around the world

Table 6.1

Quantities of waste generated as a result of one years operation of a typical 1000 MWe nuclear power plant

	Approx. weight in metric tonne	Approx. volume in cubic metres	Total activity in Curies	Notes
Uranium mill tailings		60 000	600	Reduced to 40 000 m^3 if Pu recycled in thermal reactors and to about 1000 m^3 if fast reactors used
Power station operations		200–500	3000	After concentration by filtration and ion exchange
EITHER Spent fuel	30		300×10^6	Containing about 200 kg of plutonium
OR from reprocessing of spent fuel High level wastes	0.5	3	150×10^6	After solidification from about 15 m^3. Contains 1–2 kg of residual plutonium
Cladding hulls		3	1.5×10^6	Could contain traces of plutonium
Other low and medium level wastes		100	10 000	After concentration by evaporation, filtration, ion exchange etc.

ranges from around 0.05 per cent to 1 or 2 per cent. The ore is usually extracted in the form of rocks which are crushed and treated with acid solvents to remove the uranium metal. After treatment this leaves very fine sand-like tailings which, by the very nature of the process, have had most of the radioactive uranium removed. But there are still likely to be residual traces of uranium and associated radioactive daughter products formed by the decay of uranium previously in the ore. Notable among these is a gas called radon which would normally be trapped within the rock surrounding natural deposits of uranium, but can escape more easily from tailings. This is quite harmless in the open air when the radon gas

Table 6.2

Quantities of waste generated as a result of one years operation of a typical 1000 MWe coal-fired power plant

	Metric tonnes
Ash retained	300 000
Sulphur retained	47 000
Emissions:	
Fly ash	2 000
Sulphur dioxide	24 000
Carbon dioxide	6 000 000
Carbon monoxide	1 000
Oxides of nitrogen	27 000
Mercury	5
Beryllium	0.4
Arsenic	5
Cadmium	0.001
Lead	0.2
Nickel	5
Also naturally occurring radioactive materials 0.03 Curies	

will be carried away by breezes, but has caused some concern in the past when tailings were inadvertently used for what appeared to be an excellent substitute for sand in building foundations. Some such buildings were found to accumulate unacceptable concentrations of radon gas if the ventilation was not very good. The solution was to improve the ventilation of the buildings and prevent future use of tailings as building materials.

Clearly, the lower the concentration of uranium in the ores the greater the quantity of tailings produced in gaining a given quantity of uranium. However, the lower grade ores are only likely to be exploited commercially if they occur near the surface and can be mined by open cast techniques. This means that it is also relatively easy to backfill the workings with the mine tailings and any overburden. Once this is done the escape of radon gas is not likely to be significantly different from that which came out of the ground before the uranium was mined from it.

It is also clear that the quantity of uranium tailings associated with a given amount of electricity finally produced by a nuclear power station is reduced by more efficient use of the uranium fuel. This means, for example, that a fast reactor which has the possibility of using uranium fifty to sixty times more efficiently, would cause a correspondingly smaller amount of tailings to be produced at the mine. In practice, of course, once a decision has been taken to open up a mine it is likely to be worked out, no matter how the material is finally used. However, the number of uranium mines needed to support world nuclear power programmes could in theory be reduced. In addition, if the uranium is used more efficiently, it can command a higher price and there should be more money to permit better mining practices in dealing with tailings.

The remaining processes in the production of fuel for nuclear power plants – the processing of uranium, enrichment and fabrication of fuel assemblies – generate very small quantities of waste. The processes used are designed to produce an extremely pure product and, since the material is also very valuable, they tend to recover every trace of uranium from the process streams. For the same reason, any rejected fuel in the final fabrication stages is recycled in scrap recovery plants and virtually nothing is wasted.

Waste arising at power stations

Although highly radioactive fission products retained inside the spent fuel removed from a nuclear power reactor are the waste materials that are usually being referred to when one hears of the so-called 'waste problem', the day-to-day generation of low and intermediate waste during the normal operation of a nuclear power plant is a more pressing practical problem. Well-established practices and tight regulations ensure that routine emissions of very low level wastes and the packaging and transport of intermediate wastes represents no hazard to the general public. But the processes involved in reducing these wastes to more manageable proportions are the ones in which the operating staff are most likely to come into contact with radioactive materials and are therefore the most likely cause of minor incidents and occupational exposure to radiation.

Routine release from a power station takes the form of gases, predominantly the chemically inert noble gases, krypton and xenon, and water that may contain some residual traces of dissolved radioactive substances. In water-cooled reactors the gases produced in the core region and dissolved in the water can come out of solution when the pressure is reduced for refuelling operations or, in the case of boiling water reactors, in the steam turbine. With gas-cooled reactors there may be small leakages of gas or slightly activated air drawn from a region around the pressure vessel. In all cases the buildings are provided with powerful ventilation equipment to maintain a reduced pressure inside and ensure that any leakage is inwards. Any gases escaping from, or released by, the plant are therefore drawn into the ventilation system, passed through high grade filters to remove any particulate matter and charcoal filters to remove any chemically active gases, and dispersed through a tall ventilation stack. In some cases, if short-lived radioactive gases are involved, a hold-up vessel or a labyrinth system of baffles can be used to postpone the discharge and ensure that the radioactivity has decayed to insignificant levels before release.

Low level liquid wastes are collected from a variety of sources around a complex nuclear power plant. These could simply be the water used to wash a piece of equipment that may have been contaminated with radioactive material. There may be small quantities of water from the main reactor coolant circuits which leak through the glands of pumps and valves and are collected in special drains. Or there may be larger volumes of coolant water that have been passed through filters and other chemical clean-up plant to remove all but minute traces of residual radioactivity. Subject to strict controls, this low level waste can be discharged into the sea or river which is being used to provide the cooling water for the turbo-generator condenser.

By their very nature, radioactive pollutants are easy to check and monitoring instruments can provide continuous measurement at concentrations of radioactivity well below the limits prescribed by safety authorities. This is often not the case with other types of industrial pollutants which may only be measured by taking samples and subjecting them to time-consuming analysis. If ventilation or liquid discharges are being passed through filtering systems it is also possible to obtain an early warning of higher-than-expected concentrations of radioactivity by placing monitors near the filters, and the flow can be shut off while the cause of the increase is investigated.

For these reasons the operators of nuclear power stations have had little difficulty in keeping routine releases well below the legally prescribed limits which themselves are usually based on conservative interpretation of the recommendations of the International Commission on Radiological Protection (I.C.R.P.). These limits are set at such a low level that a person camped out by the perimeter fence of a nuclear power station might at the very most be exposed to about 1 per cent of naturally occurring background radiation.

The main concern during normal operation of a nuclear power station is the accumulation of low and intermediate solid wastes. In particular the filters and ion exchange resins which are intended to trap and concentrate radioactive materials from coolants or ventilation air must be dealt with when the level of activity accumulated reaches a particular level. There is also a variety of other materials which may be slightly contaminated with radioactivity, such as protective clothing and gloves, tissues used to wipe contaminated surfaces, protective coverings put on the floor to collect drips during the transfer of items of equipment in and out of storage pools, etc.

'Good housekeeping' practices ensure that as far as possible different types and concentrations of waste material are carefully segregated and clearly identified so that they can be treated in the most appropriate way. In addition, the health physics staff, in carrying out their regular surveys, will check on the whereabouts and activity of any wastes. In this way it is possible to make sensible decisions to minimize exposure of operating staff handling the wastes. If, for example, it is known that most of the contamination of an item of equipment is due to a short-lived radioactive component it would make sense to leave it at the bottom of a storage pool or behind a wall of lead bricks for an appropriate period until most of the radioactivity had decayed away.

At an appropriate time the segregated wastes will be subjected to further processes to reduce their volume before being put into drums for longer-term storage or disposal. A large variety of commercially produced plant and equipment is now available for these volume reduction processes. Equipment includes special incinerators and powerful presses, or combinations of both, and can generally be relied upon to reduce the volume of a variety of waste materials by factors of five to ten times. Special efforts are also made to immobilize the waste materials when they are packed into storage drums. The methods used include incorporation of the waste in concrete or bitumen and there is interest in new polymers for this purpose. Depending on the radioactive content of the waste it may be necessary for some concrete shielding to be incorporated around the storage drums.

Depending on the particular circumstances a typical 1000 MWe nuclear power station could generate up to 2000 standard 200-litre drums of low and intermediate waste material in a year, although the trend is for the physical amounts to be reduced by steady improvement of volume reduction techniques. Operators will try to find an optimum trade-off between the cost of greater volume reduction and the cost of storing and disposing of a larger volume of lower concentration waste.

A broader issue is the optimum trade-off between the risk of slightly increased occupational exposure in the process of concentrating and isolating low level wastes against the small population risk from release and dispersion of some of these wastes. The present climate of opinion certainly favours concentration and isolation and, since the processing of these low and intermediate wastes does not impose severe financial burdens upon nuclear power plant economics, most operators are likely to go this way as the path of least resistance. It can, however, be argued that the total radiation exposure – man-Sieverts – to occupational staff will, as a result, be somewhat greater than the exposure to the general public if the corresponding amount of radioactive material had been widely dispersed at very low concentration into the sea or air.

Shallow land burial at carefully controlled sites is usually adequate for disposal of low level beta-gamma wastes. The level of radioactivity is such that it will have decayed to harmless levels long before any natural processes of erosion or ground water flow could release it to the environment. Deeper burial in stable geological formations is preferred for intermediate level wastes and also for any wastes that may contain traces of alpha-emitting isotopes.

A particularly impressive demonstration of disposal of both large volumes of low level wastes and smaller amounts of intermediate and alpha-emitting wastes, has been undertaken in Germany at a disused salt mine. In previous mining of salt from a gigantic dome-shaped formation, around 150 large underground chambers were created at depths down to 750 metres. Since starting to store waste in 1967, around 20 000 drums of low level waste have been safely deposited, filling only two-and-a-half of the chambers, and several thousand drums of intermediate level waste have been deposited in a further chamber. This particular salt mine is in one of several hundred similar formations in northern Europe and investigations are being carried out in a larger salt dome to assess its suitability for a purpose-built repository for high, intermediate and low level activity.

Another approach to the disposal of low and some intermediate level wastes is deep ocean dumping. Although this tends to stimulate a hostile initial reaction from many people, due to the wanton pollution of the oceans with oil and other man-made wastes, it is, in the case of nuclear waste, a responsible course of action covered by an international convention. The waste is carefully packaged in drums, again following international regulations, such that the drums will sink to the bottom of the ocean at a depth of 4000 metres. The activity that might escape from drums which are eventually corroded away by the seawater will be subjected to a gigantic dilution by the volume of water such that it will be far below the concentration which is safe to drink if it ever gets anywhere near the surface of the ocean. It is also worth noting that the oceans contain natural radioactivity from traces of uranium and associated radioactive decay products. This radioactivity amounts to far more than anything that might arise from disposal of low and intermediate level man-made wastes. Every effort has also been devoted to estimating possible pathways through the food chains of living organisms in the oceans which might conceivably reconcentrate radioactivity. But, even allowing for large uncertainties, there is no identifiable route for any significant amount of radioactivity to get back to man.

High level wastes

By far the largest quantity of radioactivity – though still, as has been mentioned, the smallest physical volume – produced during the operation of a nuclear reactor is locked firmly inside the fuel and sealed inside the cladding. This has been designed and tested to give the highest possible standard of leak tightness over several years of operation at high temperature in the core of the reactor.

If nothing more is done to spent fuel when it is unloaded from a reactor then the spent fuel itself becomes high level waste, but in fact less than 4 per cent by weight of the spent fuel is actually radioactive waste material. The vast bulk of the spent fuel is unused uranium which, if it was enriched to 3 or 4 per cent before going into the reactor, would still have residual enrichment of over 1 per cent. It is therefore well worth considering recycling this uranium rather than treating it as waste material. Also, the spent fuel contains around 1 per cent by weight of plutonium which, if treated as a waste product, is one of the more difficult radioactive materials to dispose of because of its long half-life and alpha emission. If recycled, it is a valuable fuel which can be 'burnt' either in thermal reactors or, more profitably, in fast reactors.

It is possible therefore to make a strong case for reprocessing spent fuel to separate the genuine waste into an even smaller volume and to recycle slightly enriched uranium and plutonium to yield more energy. On the other hand, reprocessing of spent fuel involves complex chemical processes which, while concentrating the high level waste, will also generate some more low and intermediate wastes of somewhat greater volume. Detailed studies in a number of countries and the deliberations of the International Nuclear Fuel Cycle Evaluation (I.N.F.C.E.) have concluded that, from a waste management point of view, either recycling or once-through options can be adopted without creating difficult problems for final safe disposal of high level wastes. The choice between the two options may ultimately be decided by political or economic considerations.

Whichever fuel cycle option is finally adopted, it makes sense to store spent fuel of the type with stainless steel or zirconium alloy cladding for some years before either reprocessing it or disposing of it. This is because the fuel cladding has been designed to withstand such severe conditions while operating in the reactor that subsequent storage at the bottom of a deep pool of water is possible for ten or twenty years without any likelihood of deterioration. During this storage period a lot of the short-lived radioactivity will decay within the first year and the later handling, whether for reprocessing or conditioning prior to final disposal, will be easier and present less risk of occupational exposure to the workers involved.

There is quite a lot of storage space for spent fuel provided in unloading ponds located at nuclear power stations. In the latest plants it is common to provide capacity for spent fuel generated over ten or more years of operation at the power station site. But with delays in reaching political decisions on whether or not to reprocess and a likely shortage of commercial reprocessing capacity for a number of years even when decisions are taken, the need has arisen in a number of countries for additional facilities for interim storage of spent fuel from a number of power stations. The first such interim store is being built in Sweden and will consist of a large pool of water in an underground rock cavern adjacent to one of the country's nuclear power stations. Plans are also well advanced for one or more interim stores in Germany and in this case a new concept of storing the fuel in a dry state inside cast-iron flasks which can also be used for transporting safely to and from the storage site, is being adopted. In Britain and France, where there are plans to provide new commercial reprocessing plants for oxide fuel, the first stage of construction, which consists of the building of large storage pools for reception of spent fuel, is well advanced. At these plants, it will be possible to accept spent fuel from customers for reprocessing services up to ten years before the fuel is actually reprocessed.

While these interim storage arrangements are safe and sensible for spent fuel clad in stainless steel or zirconium alloys, it is not possible with the earlier natural uranium fuel clad in magnesium-based alloy – Magnox – which is discharged from first generation gas-cooled reactors. This is because the cladding is attacked slowly by water and the fuel can only be stored for a year or two in ponds. As a result, Britain and France have for many years been reprocessing spent fuel on a routine basis from their gas-cooled reactor power stations and similar power stations built in Japan, Italy and Spain are doing the same. It is mainly from these reprocessing operations, together with military production of plutonium in the United States as well as Britain and France, that experience has been gained in the management of the highly concentrated wastes separated from uranium and plutonium during reprocessing.

The high level reprocessing wastes are initially in liquid form in a strong nitric acid solution. It makes good sense to store these waste liquors for a further period of five to ten years before further conditioning to immobilize them. In this period the shorter-lived contents of the waste will decay away and heat generated by radioactivity will be greatly reduced. This eases subsequent processing.

High level liquid wastes are stored in large double-walled stainless steel tanks with sensitive detection systems to provide early warning in the unlikely event of leakage into the space between the walls of the tank. Adequate standby tanks to accept the contents of any tank that does leak are also provided. The tanks themselves are contained in large concrete vaults which provide radiation shielding and a further barrier against possible escape. A multitude of cooling coils inside the tanks is provided to remove the heat, and here again there is plenty of back-up in the system to ensure a very high degree of cooling reliability. But even if cooling were to be lost it would take ten to twenty hours before the temperature rose to the point where there might be some boiling. This means that there is a relatively long period for operators to repair the fault, transfer the contents to another tank or, as a last resort, to connect fire hoses to the cooling circuits. Needless to say, the vaults containing storage tanks of this kind are also equipped with air extraction and high grade filtration plants to capture any radioactive vapours that might escape.

Storage of high level wastes in tanks of this kind requires continuous supervision and the tanks themselves, with their associated cooling and ventilation systems, are very expensive. But to put this waste management task into perspective it is again necessary to emphasize the very small volumes involved. In Britain, for example, all the high level waste produced over a period of thirty years from reprocessing of spent fuel from both civil and military reactors is held in sixteen tanks. If no other provisions were made until the end of the century there would only be a need for the addition of about one tank per year. The total space of the buildings containing the tanks and supporting services might then occupy the space of a single football pitch. The cost of such storage is a minute fraction of the cost of electricity produced by the power stations it supports.

Solidification of high level wastes

Ultimately, the objective of all high level waste management is to fix the material, whether it be very concentrated reprocessing waste or spent fuel in which the wastes are contained in a larger volume of uranium, in a highly stable solid form for ultimate safe disposal. A great deal of work has been done in many countries on techniques for solidification of liquid reprocessing wastes and, with more recent interest in the possibility of final disposal of spent fuel, work is also being done in a number of countries on techniques for compaction and encapsulation of complete fuel assemblies.

The medium which has been most thoroughly investigated for incorporation of wastes is glass – a large solid block of borosilicate glass which can be cast at a moderate temperature along with a high concentration of waste residues to form a very stable homogenous mass which is very resistant to leaching by water (see Plate 3). Other types of glass and other materials have also been looked at and some of these look quite promising, but since borosilicate glass has been studied far more extensively and looks as if it will be perfectly satisfactory, this medium is likely to be preferred at least for early solidification plants.

A variety of different plants has been developed for the incorporation of wastes into glass – known as *vitrification*. The most proven method is a continuous process which has been in use routinely at a reprocessing plant at Marcoule in the South of France since 1978 and is currently being adopted at another site in France and at plants in Britain and Germany. In this process the concentrated liquid wastes from reprocessing are first reduced to a solid residue by passing them through a rotating drum inside an eletrically heated furnace. This is known as a *calciner*. The solid waste residue is mixed with the beads of glass-making material and falls through a vertical chamber heated to a high temperature – $1150°C$ – by an electric induction heater around the outside. A stream of molten glass comes out of the bottom of the chamber and is collected in stainless steel containers, about the size of milk churns and rather similar in appearance, where it cools and solidifies as a solid block of glass.

All these processes are quite straightforward in themselves but, because the material being processed is highly radioactive, everything must be done remotely inside a massive concrete cave. The plant must also be designed for remote maintenance using robot-type manipulators and can only be observed through very thick glass windows in the cave. Any gases or vapours produced in the calciner must also be collected, and air extracted from the cave must be carefully filtered and monitored before it is released. The bulk of these gases and vapours is trapped by passing them through a counter-flow of cool water with the resultant solution being recycled into the calciner. But a combination of condensers, absorption columns and washing processes is used to trap any residual radioactive gases. This will result in the production of some new lower-activity liquid wastes which are recycled to the adjoining reprocessing plant to be treated with other low and intermediate wastes. After re-concentration the bulk of this will eventually find its way back into the vitrification plant.

The demonstration vitrification plant at Marcoule has since 1978 been casting 350-kilogram blocks of glass in 150-litre containers at the rate of about one per day. Typically, the high level reprocessing waste resulting from one year's operation of a large 1000 MWe nuclear power station will be reduced to just fifteen of these blocks. After the containers have been filled, stainless steel lids are welded on remotely and traces of contamination are cleaned off the outside. The containers are then transferred, using a special machine, to storage pits in an adjoining building. The storage facility is in a very modest-sized building with a floor area of about 22×16 metres but it has room for 220 storage pits, each capable of taking a stack of ten containers of glass. This amounts to a total capacity equal to ten years' output from the vitrification plant or the waste from almost ten years' operation of fifteen large power stations (see Plate 4).

A huge amount of detailed laboratory work has been carried out to assess the ability of glass to retain incorporated nuclear wastes and the stability of the glass itself when subjected to intense radiation for very long periods. Various techniques are available for simulating the effects of radiation over many hundreds of years and the conclusion is that most of the types of glass tested will show no serious deterioration over such extended periods.

Measurements have also been made of the rate of leaching of materials with the same chemical composition as waste from the glass. These are usually carried out with a continuous flow of distilled water over the surface of samples of glass – a very severe test likely to produce the maximum rate of leaching. The very pessimistic leach rates obtained from such tests suggest that it would take about 3000 years for a continuous flow of pure water to dissolve a typical block of vitrified waste. In practice, of course, it is inconceivable that a block of waste would be put in a repository where it would be exposed to a continuous flow of pure water. Rather it is likely to be sealed in a stable rock formation where the only water likely to come into contact with it would already be saturated with dissolved materials from the rocks themselves. A more realistic comparison has been obtained by subjecting rocks of the type that might surround the glass blocks in a final repository to the same leaching tests with a flow of pure water. These indicate that the leach rate of borosilicate glass is comparable with samples of rock which is 100 million years old and only ten to fifty times greater than rock which is 600 million years old.

4 Storage facility at a high level waste vitrification plant in France

5 Fast reactor fuel assembly

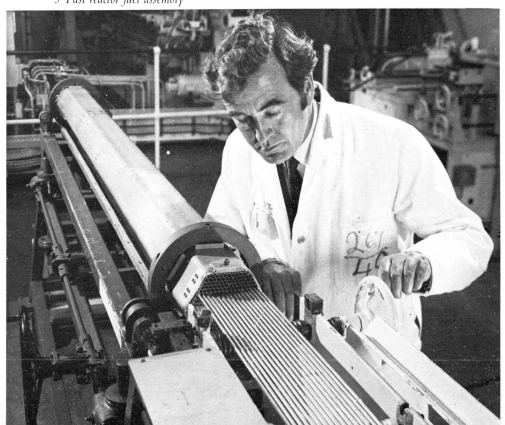

6 Conceptual tokamak power reactor

The final resting place

High level liquid waste has been stored safely in double-walled tanks for tens of years, spent fuel has been stored in ponds for more than ten years and there is no reason to believe that this period could not be extended to several decades. It should not be difficult to build a facility, similar to that at Marcoule in France, which would be suitable for the safe storage of blocks of vitrified waste for a hundred or more years. But there is still a desire to find a final resting place for nuclear waste in which it can be left without any kind of human supervision for thousands of years, and where there is no possibility of radioactive material finding its way back to the human environment. A tremendous amount of effort is being devoted to studies to show that this can indeed be done although it should be recognized that no such demanding requirements are placed on any other of man's wastes, even though they may be potentially just as toxic, far more abundant and of infinite life.

There is, of course, no way of proving anything for thousands of years. But it is possible to devise a number of barriers, each of which offers isolation on this sort of timescale, and to show that the probability of all these barriers being breached is so remote that one can claim an effective guarantee of absolute safety for radioactive waste disposal. Most of the schemes proposed for final disposal of high level radioactive waste involve deep burial in stable geological formations. But it is important to recognize that these are not relying just upon the burial for isolation of the wastes. A simplified representation of the possible barriers in a typical arrangement is shown in Figure 6.1.

The first barrier is provided by conditioning of the waste in a highly stable and insoluble form. As described above, vitrification offers a form in which it would take thousands of years to leach out the waste even if ground water were ever to reach the surface of the glass. Next there will be an additional corrosion-resistant container around the stainless steel container in which the vitrified waste is cast. This is likely to be made of thick cast-iron, although more exotic materials such as bronze, copper, lead or titanium have been suggested in some proposals. The trouble with these alternatives, apart from the high cost, is that ultimately the toxicity of the concentration of metal may represent a marginally greater risk than the radioactive waste it contains. Either way, the container offers a barrier to prevent ground water reaching the surface of the glass which should be sound for hundreds of years at the very least and possibly for a thousand or more years. The important thing, however, about the container is that it would certainly prevent any water from reaching the surface of the glass in the early years when the concentration

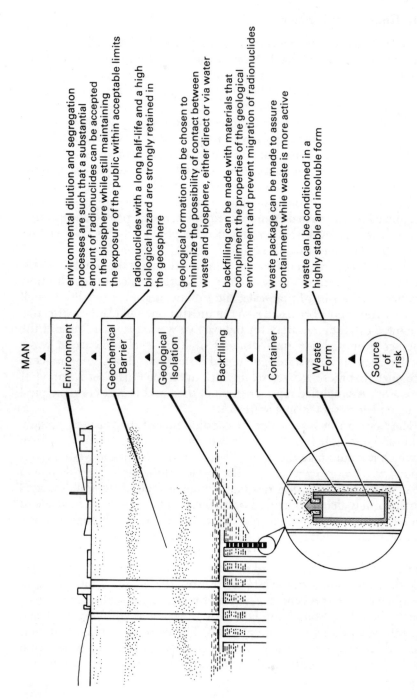

Figure 6.1 Multiple barriers for the final isolation of radioactive waste

of leachable radioactivity near the surface is still high. Later in life the radioactivity would have decayed to a much lower level and it would be necessary to dissolve a substantial amount of the highly insoluble glass to release significant concentrations of radioactivity.

The next barrier is backfilling around the containers of radioactive waste. Most repository concepts involve the placing of containers in boreholes in mined underground galleris. Once in position a backfilling would be used to seal the containers in place. The backfilling could be concrete but an alternative is a rather remarkable clay material called bentonite which is plastic enough to fill all spaces around the containers and any cracks that might occur in the rocks around them. It is highly impermeable to water but also has powerful chemical absorption properties which would cause it to trap many of the radioactive substances contained in the waste if they did get out by some means. The backfilling can therefore represent another barrier which is good for hundreds or, more likely, thousands of years.

The next barrier is provided by geological isolation. Investigation of suitable geological formations which have been stable for millions of years and are likely to remain so for millions more has attracted a good deal of publicity in recent years and has provided a fascinating field of study for geologists. In practice there are quite a few options from which to choose. Most interest has centred upon rock salt, particularly the huge underground dome structures found in profusion in Northern Europe; crystalline rock such as granite which occurs in a number of regions in the form of great unfractured masses; and deep beds of clay found in lowland areas or even below the floor of the oceans. Each of these types of formation has particular advantages but they all offer vast areas which are free from the effects of movements in the earth's crust and have survived many geological upheavals – such as ice ages – without showing any inclination to move or fracture. Another feature, which is not apparent from the distorted scale of Figure 6.1, is the likely size of the repositories relative to the geological formations. A repository large enough to take all the nuclear waste from a large industrial country for a period of fifty years would occupy a minute fraction of the volume of any one geological formation – truly a needle in a haystack.

Even if by some incredible means some radioactive material were to find its way out of the mass of a stable geological formation there is still a further geochemical barrier in the ground above. This is likely to extend to a depth of 1000 metres or more and provides a further massive volume through which material would have to diffuse in order to reach the surface environment. In the process most constituents of any escaping wastes, in particular some of the long-lived radionuclides such as plutonium, would be chemically captured in the ground.

Finally, if there is still some remote means for radioactive material to reach the surface, the natural environment offers a further barrier – or more precisely a huge dispersion volume – which is likely to protect man from exposure to any hazard.

This simple description of the available protection barriers should give an impression of the huge safety margin that can be provided when we finally dispose of radioactive waste but more searching studies, based on highly pessimistic assumptions, have been undertaken in a number of countries to investigate all possible pathways for radioactivity to get back to the human environment. Some results from just two of these are worth mentioning. Although these studies have been complemented or superseded by many others and all acknowledge the huge uncertainties in considering anything on the timescale of thousands of years, they do provide dramatic illustrations of the fact that there are equally huge factors of conservatism in estimating possible future risks from waste disposal.

A Swedish study concluded that the worst case might occur if somebody happened to drill for water in the precise location of a repository. If the water recovered were to be used directly for drinking the maximum additional radiation dose to which the people drinking it might be exposed was estimated to be almost forty times less than the dose limits recommended by the International Commission on Radiological Protection (I.C.R.P.) and about one third of the level already found in some Swedish drinking water due to natural radioactivity. This maximum dose would, incidentally, be unlikely to arise in less than 200 000 years' time because of retention of the more hazardous substances by the buffer materials and rocks.

Another study started from the improbable assumption that all the world's nuclear waste produced by the year 2000 – in some 72 000 containers – was, simply, to be dumped at one location on the ocean bed in mid-Atlantic. (It must be emphasized that nobody is actually suggesting this solution.) It was assumed, conservatively, that the steel containers would only survive for a few years and that the blocks of vitrified waste would fragment into fist-sized lumps. From this extremely pessimistic starting point it was estimated that all the vitrified waste might be dissolved by the seawater over a period of 3500 years. Direct dilution in the ocean waters would reduce the concentration of radioactivity to far below safe drinking limits but consideration was also given to possible pathways to man through marine life. Here the worst case was estimated to arise if at sometime in the future we are so short of food that we start to recover deep ocean plankton for direct human consumption. This might expose the people concerned to about one hundredth of the I.C.R.P. limits.

Waste Management

This last, almost farcical, case study illustrates the great potential barrier offered by the ocean depths. While most work is currently concentrated on the investigation of land-based locations for waste repositories, there is a strong body of scientific opinion which believes that ocean bed burial – not just dumping on the ocean floor as considered in the ultra-pessimistic study – might offer a better alternative even if instinctive public reaction is more hostile. Another suggestion, also based on the potential of a large ocean mass to dilute any escaping activity, is for the location of repositores below the surface of remote islands.

Timescale and cost

The conceptual nature of various proposals for ultimate waste disposal frequently raises the question of why nothing is yet being done to provide actual repositories. The extreme nature of some of the isolation schemes proposed also raises questions about the likely cost and even stimulates the belief that if such measures are being considered the 'stuff' must be really nasty.

In reality it makes a lot of sense to delay final disposal and to keep open the many options in the hope that further studies will lead to the optimum choice. The reason for delaying final disposal is related to the heat generated by radioactive waste. When first vitrified, high level wastes can generate around 10 Watts/litre within the mass of the glass block. This heat can be removed by a simple flow of cooling air provided by fans or even natural circulation in a properly designed storage facility, but if the waste was embedded in rock, the local temperature could rise to levels which might encourage fracturing of the surrounding rock and increase the rate of leaching by any ground water. After periods of thirty to fifty years, however, the radioactivity will have decayed to such a level that vitrified blocks can be embedded close together in a final repository without surface temperatures rising above $100°C$ to $200°C$, or not significantly higher than natural temperatures occurring in deep rock formations. Most schemes therefore envisage a period of storage in surface or shallow underground interim stores which can be carefully supervized and cooled. A further advantage of interim storage is reduction in the level of surface emission of radiation which simplifies the final disposal operations and will reduce the occupational exposure of the workers involved.

The apparently very elaborate and costly proposals for disposal of nuclear waste certainly represent a response to the fact that the public has been led to believe that there is a high risk. In reality, however, the cost will be very modest in relation to the value of the electricity produced by the nuclear power stations generating the waste. Conservative estimates on a number of conceptual schemes for waste disposal seem to agree on a cost of around 1 per cent of the electricity production cost. In several countries operators of nuclear power stations are already setting aside money in special funds which should be able to meet the waste management costs very adequately.

One reason for not being in too much of a rush to decide on final isolation methods for the waste from nuclear power plants is the possibility of developing more sophisticated techniques for segregation of the different constituents of the waste so that each can be treated in the most appropriate manner. Mention was made in Chapter 1 (page 14) of the way in which the different isotopes found in a mixture of radioactive waste will decay at different rates, with the short-lived isotopes contributing most to the intense radiation emission in the early years and the longer-lived isotopes taking over in the longer term at a lower level of radiation emission. This rather complex situation is represented in Figure 6.2 which shows decay curves for the main constituents of the high level waste that would result from the reprocessing of spent fuel from one year's operation of a light water reactor.

Bearing in mind the logarithmic scale in which every division on the graph represents a factor of ten, it is possible to identify the main contribution to the initially very high level of radioactivity as relatively short-lived fission products – notably cesium-137 and strontium-90. But after 100 or 200 years the contribution from these isotopes decays to insignificant levels compared with longer-lived isotopes, most of which are in a family of elements known as *actinides*. These elements do not occur naturally but are formed, like plutonium, by successive capture of neutrons in the nuclei of uranium atoms and the subsequent product atoms. The short-lived fission products predominantly emit beta and gamma radiation when they decay while many of the long-lived actinides are alpha emitters.

If the actinides could be chemically separated from the bulk of fission product wastes it would be possible to treat the batches rather differently. Protection against the very high level of radioactivity of the fission products would have to be provided for around 200 years but after that they would be harmless and it would not be necessary to go to the extreme of geological disposal already described. The much smaller quantities of actinides, on the other hand, would be relatively easy to handle because from the outset their activity is about one hundredth of the fission product activity and only some ten times the radioactive content of the uranium used to fuel the reactor in the first place. But because they are long lived and contain alpha emitters the more extreme measures of geological isolation from the environment are desirable.

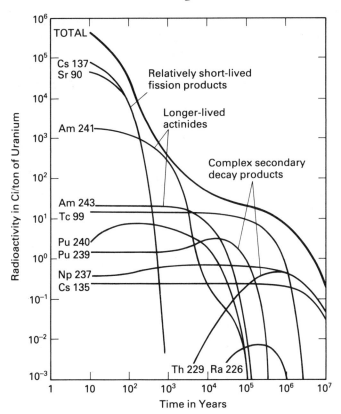

Figure 6.2 Complex composition of the decay curve for the waste from a typical reactor

Separation of actinides might open up a further alternative referred to as *incineration*. If irradiated for a long enough time with high energy neutrons, the atoms would be gradually broken down by fission into short-lived fragments that could be treated like other fission product wastes. The obvious place to do this would be in a fast reactor. Some experiments along these lines are being conducted as a joint trans-Atlantic project, with an American team undertaking the difficult task of chemical separation of very small quantities of actinides from larger volumes of fission product waste, and Britain providing facilities for irradiation in the Prototype Fast Reactor.

A further method for disposal of separated actinides that has been suggested would be to shoot them out of the earth's orbit and eventually into the sun. While this very extreme measure tends to stimulate an initial reaction of horror, due to the spectre of crashing rockets, it would not be quite as irresponsible as it sounds. If individual consignments were kept sufficiently small it would be possible to ensure that the worst possible failure of one rocket would distribute levels of radioactivity into the stratosphere which were far below those resulting from the early atmospheric testing of a single nuclear weapon. And while nobody wants to see such releases it is very unlikely that any harm to world populations would result.

The bigger question about actinide separation is whether it is really justified in the overall reduction of statistical risk. The argument is roughly this: While separation of actinides might eliminate the already very small risk that in a thousand years' time a population of around one million might each be exposed to one millionth of the radiation dose limits, there is a significantly higher, though still small, risk that today one worker would be exposed to the full radiation dose limit during the operations of separation. It is impossible to put precise figures into such an argument but it is possible to see how the overall risk of somebody being harmed could well be increased if we take what appear to be more extreme protective measures.

Decommissioning of power stations

With a number of early commercial nuclear power stations already nearing the end of their original design life, and with a number of small experimental and demonstration reactors already closed down, it is important to consider what should finally happen to the structure of the reactors which, over their lifetime, will have become contaminated to an appreciable level with radioactivity. A number of detailed studies of decommissioning have been undertaken and some of the closed-down experimental reactors will be used to demonstrate the techniques.

Once the fuel has been unloaded and taken away for reprocessing or direct conditioning as high level waste, the main task is to deal with a considerable volume of structural steelwork in and around the reactor core. During operation of the reactor this is irradiated by neutrons and becomes radioactive, mainly due to cobalt in the steel. The level of radioactivity is nothing like that of high level waste and there is no long-term problem because the half-lives of the predominant isotopes are measured in tens of years. But the cobalt isotope emits penetrating gamma radiation and there might therefore be some problems of occupational exposure if workmen were asked to cut it up immediately after the plant is closed down.

The most sensible course of action, which is contained in most proposals for decommissioning, will therefore be to clear all the conventional plant and buildings around the reactor but to leave the relatively small volume of the reactor vessel and surrounding concrete shielding intact for a period of thirty to fifty years. This would allow radioactivity inside to decay away harmlessly to levels at which it can be handled relatively easily. At this stage too it would be perfectly safe to dispose of cut-up steelwork by shallow land burial or ocean dumping. It has even been suggested that, if by that time we are running into shortages of steel, it would be possible to recycle material for use in new reactor pressure vessels or pipework for nuclear plants.

This course of action is consistent with land use of the site of a nuclear power plant. It is most likely that, as one plant is taken out of service, there would be a new plant and the sensible place to do this would be at an existing site with cooling water installations, roads and transmission lines. The existence of one or two concrete blocks containing the old reactors would only occupy a small portion of the site and need not get in the way of new plant construction. At the same time the organization for supervision of the site would be available to keep a check on the old reactors during the cooling-off period and supervize the final disassembly at an appropriate time.

The same general principles apply to other nuclear facilities such as reprocessing plants although the detailed problems will be different. Notably at a reprocessing plant, there is a need to decontaminate the surfaces that have become contaminated with the highly radioactive substances which have been processed. If there is a need to reuse the same buildings, decontamination can be carried out with a variety of chemical treatments shortly after the plant has been taken out of service. This has in fact been done at early reprocessing plants in both Belgium and Britain, where it was possible to enter processing cells to take out old plant and make way for new improved equipment.

7

Advanced Reactors

The first type of reactor to be used to drive a small generator as a simple demonstration of the potential of nuclear energy to produce electricity, was a fast reactor – the Experimental Breeder Reactor I (E.B.R. I) – in the United States as early as 1954. This is an indication that while more powerful thermal reactors had been operated for military production of plutonium and were being developed for submarine propulsion, the feeling at the time was that the only way in which nuclear energy could make a significant contribution to the much larger civil energy requirements would be through the development of more advanced reactors which could use uranium resources more efficiently than the early thermal reactors. The fast reactor, with its potential to breed slightly more plutonium fuel than is being consumed in the core of the reactor, was the most promising of the advanced reactors but a number of different concepts – notably systems referred to as *advanced converters* because of their ability to convert the metal thorium into a new, and very efficient fuel, uranium-233 – were studied, and several have been developed to the stage of experimental and demonstration reactors.

Subsequently a number of factors changed the priority for civil nuclear development to thermal reactors. Most important of these factors was the discovery of much more commercially exploitable uranium around the world and a sharp decline in the military requirement for uranium. The momentum of thermal reactor development for military purposes also provided the civil nuclear industry with a strong base from which to develop large commercial reactors. These could compete in electricity production with traditional fossil fuels. But the leading industrial countries of the world have still maintained a large parallel development effort on the fast reactor because of its long-term potential to provide a virtually unlimited source of energy and the political attraction

Advanced Reactors

of a system which offers a high degree of national self-sufficiency in fuel supplies.

A sizable development effort has also been maintained in a number of countries on the other advanced converter reactors which can use thorium as the feed material to produce fuel. Most important of these is the high temperature gas-cooled reactor (referred to as H.T.G.R. or H.T.R.) which has attractive possibilities for producing process heat for a variety of industrial applications as well as electricity. Heavy water reactors, such as the Candu type, also offer a commercially proven system which could fairly easily be adapted to use the thorium/uranium-233 fuel cycle with a good conversion efficiency. It should also be noted that fast reactors can convert thorium to uranium-233 more efficiently than the other advanced reactors.

Fast reactors

As was explained in Chapter 2, fast reactors are so called because the neutrons emitted at high speed by fissioning atoms of uranium are used to sustain the fission chain reactor without being slowed down first in moderating material. This requires a higher concentration of the fissile atoms – either uranium-235 or plutonium – in the core of the reactor and a greater intensity (or flux) of neutrons. In this way, although the probability of a neutron scoring a direct hit and fissioning another uranium atom is less with fast neutrons than it is with slow (thermal) neutrons, it can be raised to a sufficiently high level to get a chain reaction in a fast reactor due to the greater density of fissile atoms. What is important to recognize is that this does not imply a reactor which is faster and more difficult to control than a thermal reactor – the rate at which atoms must be split to produce a given output of power is the same in a fast reactor as it is in a thermal reactor.

Nor should the name fast be linked to the breeding possibilities of the reactor. A fast reactor can breed more plutonium fuel in the blanket of depleted uranium around the core than it is consuming in the core but in no sense is this a fast process. In future commercial fast reactors one might expect the blanket elements to stay in place around the reactor core for several years, until they have accumulated enough plutonium to justify their removal for reprocessing. And it could take twelve to fifteen years to recover enough additional plutonium to provide the initial feed of fuel for a new fast reactor. Because of the misunderstanding on this point and the spectre promoted by opponents of a 'plutonium economy' in which there is runaway production of plutonium around the world, there has been an attempt to decouple the words 'fast' and 'breeder' in talking about these reactors although, to confuse matters, Europeans are tending to refer to 'fast reactors' while Americans prefer 'breeder reactors'.

Without any moderator, the fast reactor does, however, have a very compact core producing a lot of heat – referred to as a *high power density* – and one of the prime considerations is the provision of a very efficient coolant to remove this heat from the core and transfer it to steam-producing plant. As with thermal reactor development, a number of different options were considered for the cooling of fast reactors in the early days – including sodium, helium gas, and steam – but over the years the fast reactor development programmes in virtually all countries have tended to concentrate on sodium. There is still an enthusiastic body of opinion that supports certain virtues of gas cooling for fast reactors, but adherents to this view would have to admit that the status of technological development is now far behind sodium-cooled systems and could only catch up with the help of a very large development budget which at present does not seem likely to be forthcoming.

The state of development of sodium-cooled fast reactors is, on the other hand, ready for commercial introduction as and when a reasonable economic case can be made. Large prototype power plants, with output capacities which are comparable with or larger than many of the early commercial thermal power stations, have now been operating for several years in France, Britain and the Soviet Union. Similar plants are under construction or planned in Germany, the United States and Japan but these have suffered badly from politically-motivated postponements and licensing delays. The first fast reactor power station which could be classed as a demonstration of a commercial-scale plant is a 600 MWe reactor at Beloyarsk in the Soviet Union which was put into operation during 1981, but the largest commercial-sized demonstration is the Super-Phenix reactor with a capacity of 1200 MWe which is nearing completion at Creys-Malville in France and is due to be put into operation around 1984. A detailed design has been developed for a commercial demonstration fast reactor (C.D.F.R.) in Britain which is comparable in size to the French plant but a final decision to proceed cannot be taken before a public inquiry which is unlikely to take place before 1985.

The truly commercial introduction of fast reactors will come with series ordering of standard units which are better able to compete with current thermal reactor power stations by virtue of replication. The Soviet Union has started detailed design work on an 800 MWe unit which could be the first of a series of commercial units starting to come into service in the early 1990s. The French are also working on a Super-Phenix II design with a capacity of 1500 MWe and hope to get political decisions on whether or not to proceed with series ordering of four to six such units after about a year of operation of Super-Phenix – this means around 1986.

France, Britain and the Soviet Union have all evolved a very similar design of fast reactor, referred to as a *pool-type* because the whole of the primary circuit is contained in a single large pool of sodium which itself is contained in a large round-bottomed vessel with a diameter of around 20 metres (Figure 7.1). One of the advantages of the use of sodium as a coolant is that its boiling temperature is well above the 500°C to 600°C operating range of the reactor and there is therefore no need for a reactor vessel of massive construction in order to pressurize the coolant. Inside the vessel the sodium is divided into two regions: 'hot' sodium above the reactor core and 'cold' sodium around and below the core. Circulation pumps suspended from the top of the vessel force cold sodium into the bottom of the reactor core and up through the fuel assemblies to emerge in the hot sodium region above the core. The sodium is then directed outwards to a number of intermediate heat exchangers which are situated around the core and suspended, like the pumps, from the top of the vessel. The sodium passes down through these intermediate heat exchangers and gives up the heat acquired from the core to a secondary coolant. It then emerges from the intermediate heat exchangers into the cold sodium region for recirculation.

Because the intense flux of neutrons in the core of a fast reactor induces a relatively high level of radioactivity in the sodium of the primary coolant as it passes through the core and because, as is well known, sodium reacts violently if it comes into contact with water, it is considered expedient in such plants to have a secondary buffer coolant between the active primary coolant and the steam-rising section of the plant. This secondary circuit also uses sodium coolant but, because it has not passed through the core region, it is not radioactive. It is pumped through the tubes of the intermediate heat exchangers in the main vessel and then transfers the energy, via a number of circuits external to the vessel, to steam generator units. It is, of course, still very important to avoid any sizable leaks in the tubing of the steam generators because of the damage that could be caused by the reaction of sodium and water, but the absence of radioactivity from secondary sodium makes this a less severe potential accident. In practice, very sensitive leak detection instruments have been developed and, with one exception in the Soviet Union, it has been possible to isolate steam generator units long before any leaks develop to the point where the sodium/water reaction could cause damage.

The alternative design of the sodium-cooled fast reactor, referred to as

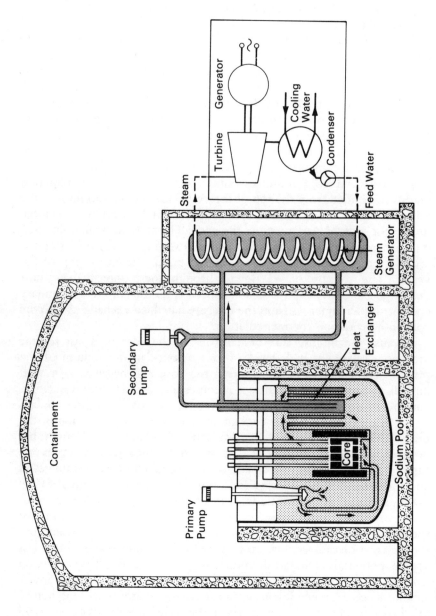

Figure 7.1 Fast reactor

Advanced Reactors

a *loop design*, differs from the pool design in that the primary circulation pumps and intermediate heat exchangers are incorporated in a number of separate circuits outside a main vessel which contains just the reactor core. In this respect the configuration is more like a pressurized water reactor although there is still a need for secondary sodium circuits to transfer heat to steam generators. While the loop design introduces the hypothetical possibilities of pipe breaks and leaks in the primary circuit, it does, on the other hand, reduce the amount of sodium needed in the primary circuit and provides for greater segregation of components such as the primary pumps and intermediate heat exchangers. On balance there is not a lot to choose between the two concepts but development programmes in the United States, Germany and Japan, where the loop design has been adopted, have, for a variety of reasons, fallen behind the programmes based on the pool design. The pool design is therefore ahead at this time but the general feeling is that there is a lot of technology that is common to both designs and therefore the parallel development programmes are likely to be mutually beneficial. At a later stage of commercial exploitation we may see two competing designs of fast reactor or, alternatively, it would be fairly easy for countries to switch from one to the other without wasting too much of the national development effort.

Fuel for a fast reactor must be enriched to above 20 per cent with one of the easily fissionable atoms: uranium-235, present to a level of 0.7 per cent in natural uranium; plutonium-239, bred from uranium 238; or uranium-233, bred from thorium. Although some early fast reactors were started up with enriched uranium fuel, the main intention is to use them to consume the plutonium produced initially in thermal reactors. Plutonium separated from spent fuel in reprocessing plants is mixed, usually in the form of oxide powder, with natural or depleted uranium oxide to produce the necessary concentration of fissionable atoms. The mixed oxide powder is sintered into ceramic pellets similar to, but somewhat smaller than, thermal reactor fuel pellets. The pellets are sealed into stainless steel cladding tubes to form fuel elements – often referred to as *pins* – with an active length of about 1 metre. These fuel pins are mounted in carefully spaced bundles in hexagonal steel sheaths to form fuel assemblies which can be loaded into the reactor core (see Plate 5).

To obtain breeding of plutonium it is necessary to provide a blanket of depleted uranium around the core. This is done with rows of breeder assemblies, similar in general construction to the fuel assemblies, loaded around the core and also with extensions containing depleted uranium pellets in the top and bottom of the main fuel pins to provide a radial blanket. Breeding also takes place in the uranium of the core itself providing some *in situ* replacement of plutonium fuel and extending the useful life of the fuel assemblies in the core.

Fuel is loaded into and withdrawn from the core of the reactor about twice a year, with a straight-lifting machine which is mounted above the vessel on double or treble rotating plugs of heavy shielding material. The rotating plugs give access to all the fuel and blanket assembly locations in and around the core and allow transfer to and from the side of the vessel. Here a separate transfer arrangement – usually an inclined chute – is used to move assemblies to and from an external storage pool of sodium through a lock above the vessel. After a suitable cooling period, spent fuel assemblies and blanket assemblies can be transferred from the external storage pool to a reprocessing plant for extraction of plutonium and recycling to the same, or another, fast reactor via a fuel refabrication plant.

The fuel cycle services needed to support a fast reactor – fuel fabrication, reprocessing and waste management – employ the same basic technology as the fuel cycle services for thermal reactors, but special techniques and procedures have to be developed to cope with particular problems presented by the relatively high plutonium content in the fuel. These are not difficult problems in themselves and most of the operations have already been developed to the stage of pilot scale plants able to support the present prototype and demonstration fast reactors.

Commercial introduction of fast reactors will, however, be more dependent on the parallel introduction of complete fuel cycle services than was the case with thermal reactors. This is because the economics of the fast reactor will be strongly influenced by the time taken to recycle valuable plutonium fuel. Clearly there is a need to minimize the inventory of plutonium tied up with the operation of a reactor at any one time. A likely target for future power plants of around 1000 MWe would be an inventory of about 3 tonnes of plutonium with one third producing power in the reactor, one third cooling down in the storage pool or in transit to the reprocessing plant and one third in various stages of reprocessing and refabrication. Such a scheme presents an operational challenge for the future. One possible solution that is widely favoured is the concept of an energy park with four or more fast reactors and dedicated reprocessing and fuel fabrication plants, all on the same site.

The case for fast reactors

Inherent safety features are evolving as one of the attractions of fast reactors. Although instinctively it was believed that fast reactors would

Advanced Reactors

present more formidable safety problems than thermal reactors they are in practice turning out to be very docile reactors to operate and exhibiting some desirable safety features. One such feature is a negative temperature coefficient of reactivity, which means that the reactor power tends to regulate itself – if the temperature rises sharply the reactor will try to shut itself down without the need to insert control rods. Of course, diverse systems of control and shut-down rods are still available to reduce the power automatically and quickly if required.

Another important feature with sodium-cooled fast reactors, especially those with the pool-type design, is the ease with which natural circulation cooling can be established. Tests have shown that, if the main circulation pumps are stopped, natural circulation of the sodium coolant can still generate sufficient flow through the reactor core to remove decay heat from the core and the large volume of sodium in the pool acts as a sizable heat sink such that, even if secondary and stand-by heat removal systems were inoperative, it could take up to twelve hours for temperatures to rise to levels that might give cause for concern. It has even been shown theoretically that, if the pumps failed and control rods were not inserted to shut the reactor down, the power level would still come down to a safe low level, due to the negative temperature coefficient and the natural circulation which would remove the heat. It may be slightly overstating the case to say that fast reactors have 'walk away' safety characteristics but at least there is plenty of time for plant operators to stop and think about the best corrective action to take if anything does go wrong.

Fast reactors do, like any other reactor type, have their own special safety problems which must be very thoroughly investigated. Not least of these is the hazardous nature of the sodium coolant. A very great deal of development work and exhaustive testing on sodium test rigs has been carried out around the world and those working in the field maintain that sodium coolant – which can operate at atmospheric pressure – presents no more problems than systems using gas, water or steam at very high pressures. In particular there is confidence in the ability to detect and deal with small leaks in sodium systems before they can develop into large leaks and there is growing confidence in the ability to contain large escapes of hot sodium from coolant circuits. Although sodium burns spontaneously in air it is a relatively non-violent form of combustion, taking place only at the surface. Special spillage collection trays have therefore been devised to reduce the surface area of any sodium collected and extinguishing powders have been developed to smother any fires.

Another cause of special concern in the case of fast reactors is the high concentration fissile material in the compact core. It is argued that, if the reactor were somehow to run out of control to such an extent that the core melts, the molten fuel could collect in the bottom of the vessel in a sufficiently large volume to form a critical mass in which the fast fission reaction would continue to produce power. This could in turn cause the molten mass to melt its way through the bottom of the vessel. While it is almost impossible to visualize how this could happen with present fast reactor designs, they are still fitted with so-called *core catchers* in the bottom of the vessels to disperse any molten fuel in such a way that it could not form a critical mass. Yet another hypothetical situation relates to a possible compaction of a fast reactor core. The core would become more reactive if compacted and, in the remote chance of this happening, it might cause a rapid increase of power and release energy with an explosive force. But while this hypothetical concern is about a compact core the same compactness makes it relatively easy to provide a containment building designed to withstand the maximum possible energy release, even though nobody can explain how such a large energy release could occur in practice.

The most difficult case to make for the fast reactor is no longer one of technology of safety but rather one of economics and timing. Largely as a result of the expediency of having a secondary sodium circuit, the capital cost of a fast reactor power station is always likely to be higher than a thermal power station of comparable capacity. This, eventually, should be compensated for by lower fuel costs due to the fact that each tonne of uranium dug out of the ground can produce about 60 times as much energy if it is used in fast reactors rather than present commercial light water reactors. But these lower fuel costs cannot be realized while there is a glut of uranium on the world market and the prices are artificially low. Nobody doubts that if nuclear power programmes expand around the world the time will come when uranium prices rise to the point where fast reactors will become increasingly competitive but it is very difficult to say exactly when that will be.

The present designs of demonstration fast reactors are expected to produce electricity at costs which are about 30 per cent higher than well-established commercial designs of thermal reactor power stations, but this is still comparable to the cost of electricity from coal-fired power stations and a good deal cheaper than electricity from oil-fired plants. Designers, particularly in France, Britain and the Soviet Union, are making strenuous efforts to get capital costs of future fast reactors down – typically to within 40 per cent of the cost of a pressurized water reactor or within 20 per cent of the cost of an advanced gas-cooled reactor. They are unlikely to do much better than this except by series building of a number of identical fast reactor power stations which would allow development and tooling costs for major component parts to be spread over the programme of construction rather than having to be recovered from one order.

While the economic case for fast reactors is likely to remain slightly on the wrong side of the balance, at least until the late 1990s or early into the next century, there is still a strong case for pressing ahead with the construction of some fast reactor units. This is because they offer such great potential for long-term security of energy supplies. This huge potential is highlighted by two examples. Domestic uranium resources in France are only sufficient to meet about half present consumption in thermal reactors but if used in fast reactors this same uranium would represent a domestic energy resource equivalent in size to the oil reserves of Saudi Arabia. Britain has no domestic production of uranium but stockpiles of depleted uranium already held in the country as a waste product from military and civil uranium enrichment would, if used as feed for fast reactors, provide energy equivalent to the country's present coal production for 400 years.

Finally, there is a moral obligation on man to us the earth's resources as efficiently as possible. The energy obtained from a tonne of uranium already compares very favourably with a tonne of coal or oil but if it is possible to get sixty times as much energy from it then we should, eventually, do so. There is also a moral obligation on man to reduce the waste products that he discharges to the environment or returns to the depths of the earth. Since fast reactors consume plutonium, which would otherwise be a waste product, from thermal reactors and only breed more plutonium according to future fuel needs, they clearly help to meet this objective.

High temperature reactors

From an early stage of development of nuclear reactors for civil power production there has been an effort to find systems that operate at higher temperatures, because the higher the temperature of a source of energy the greater is the efficiency with which heat can be transferred to where it is wanted. In the case of gas-cooled reactors this was first seen with the move from first generation plants restricted to a maximum coolant temperature of around 400°C by the relatively low melting point of the Magnox cladding, to the advanced gas-cooled reactor in which the coolant temperature can be as high as 650°C as a result of the use of stainless steel cladding. The corresponding increase in thermal efficiency of power stations – the ratio of the electrical power sent out to the thermal power generated in the reactor core – is from around 30 per cent to almost 45 per cent. The effect of this is to reduce the amount of low temperature waste heat discharged from the condenser of the turbogenerator. But, as has already been explained, this increased efficiency is obtained at the expense of enrichment of the uranium fuel to overcome the greater absorption of chain reaction neutrons in stainless steel compared with Magnox.

One way to achieve yet higher temperatures and, at the same time, reduce the loss of neutrons in fuel cladding material is to use a ceramic coating of graphite – one of the efficient neutron moderators – as a cladding material for the fuel. This is the basis of a system known as a high temperature gas-cooled reactor (H.T.G.R. or H.T.R.) which has been the subject of a considerable amount of development work but which has so far failed to make the breakthrough into commercial application.

In practice the fuel consists of small particles of uranium carbide or oxide – which are also high temperature ceramic materials – coated first with a layer of silicon carbide to provide an efficient barrier for retaining fission products and then with a further coating of graphite. These coated particles, which are about 2 mm in diameter, are incorporated in larger blocks of graphite which then constitute a combined fuel element and moderator block for loading into the core of a reactor.

Two different H.T.G.R. concepts have been developed to the stage of demonstration power plants. In one hexagonal cross-section blocks of fuel and moderator are stacked in the reactor core, with a large number of channels passing through the blocks to provide a passage for cooling gas. The gas then transfers the heat to steam generators in much the same way as other gas-cooled reactor power plants (Figure 7.2). This system was developed as a joint European project and also in the United States where a 330 MWe station is now operating. The other concept is a novel design developed in Germany which uses spherical fuel-moderator elements about 10 cm in diameter. These are simply heaped into the core of the reactor with the coolant gas circulating through the spaces between the spheres. Known as a *pebble bed reactor*, this system has operated very well as a small prototype, and a larger 300 MWe power plant is nearing completion in Germany. For a variety of reasons the H.T.G.R. has not yet been adopted for commercial power stations but it still offers two interesting possibilities which have longer-term interest and are therefore justifying continued development, albeit on a fairly modest scale.

The first of these is the suitability of the reactor for a fuel cycle involving the conversion of thorium to the fissile fuel uranium-233. With no struc-

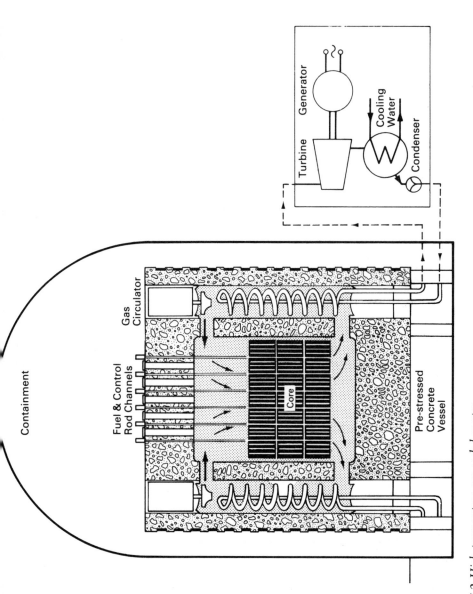

Figure 7.2 *High temperature gas-cooled reactor*

tural or cladding material in the core region other than graphite moderator, the loss of neutrons by non-productive absorption is very low and the system is said to have a good neutron economy. It is therefore possible to put thorium in the core, in the form of coated carbide particles similar to the fuel particles and also incorporated in the fuel-moderator graphite blocks. As uranium-235 atoms are consumed by the fission process they are in part replaced by uranium-233 atoms formed by the capture of spare neutrons by thorium. In effect a new fuel is bred *in situ* and contributes to the power production of the reactor, although it still becomes necessary eventually to reprocess spent fuel to remove the fission products that build up and inhibit the chain reaction. Unlike the fast reactor, it is not quite possible to reach the situation where more new fuel is being produced from thorium than the initial uranium feed but the system uses very much less uranium than other thermal reactors and also opens up the possibility of using thorium resources which are at least as abundant as uranium. The H.T.G.R. can also be operated very efficiently with the uranium-plutonium fuel cycle like other thermal reactors and therefore offers the attraction of considerable flexibility in fuel cycle strategy to suit prevailing market conditions. It should, however, be mentioned that, before this flexibility can be exploited, the technology for supporting fuel cycle services, and notably reprocessing of coated particle fuel would have to be developed on an industrial scale.

The other long-term possibility offered by the H.T.G.R. would be to make use of the very high temperature of the coolant gas. Instead of using the heat energy of this gas to produce steam and drive a steam turbine, it is possible to use it to drive a gas turbine directly. This could push thermal efficiencies up to 60 per cent or more and would greatly reduce the waste heat rejection from an electricity-producing power station. Alternatively, the high temperature energy could find other applications in a variety of chemical processes – notably the gasification of coal or steel making. Temperatures up to 1000°C are needed to drive either direct cycle gas turbines or for process heat applications. There is no doubt that such temperatures can be produced in the core of H.T.G.R.s – a small German plant has operated for extended periods with an outlet gas temperature of 950°C – but there remains some development work to be done on the plant and equipment associated with the non-nuclear parts of such plants operating at these temperatures.

Thus, the H.T.G.R. has the potential for utilizing uranium more efficiently, for exploiting thorium resources, for generating electricity more efficiently with less waste heat, and for extending nuclear energy into areas other than electricity production. These features are enough to maintain interest but at present the need to exploit them is not considered to be sufficiently pressing to justify the considerable cost of launching a

Advanced Reactors

new type of reactor and its associated fuel cycle services onto the commercial market.

Dual purpose and special purpose reactors

As well as gaining efficiency with more advanced designs of reactors, consideration has been given to more advanced ways of using the energy from any type of reactor. The most obvious way of getting more useful energy out of a power station – whether it be nuclear, coal or oil – is to use the low temperature waste heat normally rejected through the condenser of the turbine to supply district heating or other industrial users. With light water reactors, the first generation gas-cooled reactors, and heavy water reactors, the case for trying to do something useful with this waste heat is strengthened by the fact that their overall thermal efficiency is slightly lower than modern fossil-fuelled plants and therefore there is a larger amount of rejected heat for a given amount of electricity production. On the other hand the incentive to utilize waste heat from a nuclear plant is not as great because one is able to get so much cheap electrical energy from uranium fuel that is currently very abundant and not usable for anything else.

It should be recognized that the reason for rejecting heat in the condenser is to create the largest possible pressure difference between the hot steam entering the turbine and the partial vacuum created by condensing the steam leaving the turbine. It is this pressure difference that causes the steam to rush through the turbine and spin it and the electrical generator at high speed. The amount of reject heat is determined by a careful process of optimization, aimed at getting the maximum electrical energy for the given steam conditions – temperature and pressure – that can be supplied by a particular type of reactor. Any attempt to use the waste heat is liable to disrupt the optimum operating conditions and reduce electricity production, so it can only be justified if the value of the waste energy recovered is at least equal to the value of the lost electricity generation.

The difficulty to date has been in finding situations where the market for the reject heat is sufficient to compensate for reduction of electricity production. There are a number of problems. A nuclear power station generating 1000 MWe of electricity with an overall thermal efficiency of 30 per cent would produce 2000 MWe of waste heat. Even to start to use this amount of heat would take a huge interconnected district heating scheme serving a large urban conurbation in which virtually all homes and commercial premises use the district heating supply rather than alternative oil or gas boilers. Some such district heating schemes do exist in parts of northern and central Europe (including Scandinavia) but

Nuclear Energy

establishing widespread district heating in cities which currently have a great diversity of heating systems would involve massive disruption and huge expense. Since nuclear power plants have so far been located at relatively remote sites there is also a need for a long and expensive connection to district heating systems. As a result, even with rapidly rising costs of alternative systems of heating, it is difficult to find more than a handful of situations where a reasonable economic case can be made for the supply of district heating schemes from large nuclear power stations.

The possibilities for using reject heat for other industrial applications is potentially more attractive since some industrial users have processes which could consume a large block of energy and their pattern of use could be tailored more closely to complement the parallel demand pattern for electricity. But the low temperatures at which heat is usually rejected from turbine condensers would make it useless for most industrial processes. Industrial users would only be interested if they could take some of the steam from the intermediate stages of the turbine and this would result in a correspondingly larger reduction in the optimization for electricity production.

For these reasons the examples of dual purpose nuclear power plants are few and far between. The first small power reactor in Sweden was used to supply district heating to a Stockholm suburb. This plant has since been shut down but schemes for supplying district heating networks through long underground tunnels from large nuclear power stations are being seriously considered for both Stockholm and Helsinki in Finland. In the Soviet Union a prototype fast reactor supplies steam for a desalination plant which provides the water for a new town built in a desert region on the edge of the Caspian Sea, and a pressurized water reactor plant is being built to supply electricity and district heating to the city of Odessa. In Canada, one of the large nuclear power stations supplies steam for an adjoining heavy water plant and in Britain the first Magnox reactors supply some steam for the adjoining reprocessing plant. A pressurized water reactor operating in Switzerland has started supplying steam to a nearby cardboard packaging plant. The only large power plant planned from the outset as an integrated dual purpose commercial project with an adjoining chemical complex is in the American state of Michigan and has suffered from severe licensing delays which have put the start of operation back to at least 1984. Thus, while there is a large potential for nuclear energy to supply heat rather than electricity for a large number of uses which are at present dependent of oil or gas, little has so far been realized with dual purpose plants.

An alternative approach which is now gaining some interest is the use of small special-purpose reactors to supply heat only. If designed from

Advanced Reactors

the outset with heat applications in mind, such reactors can be very conservatively rated and can incorporate a great deal of inherent safety. This means that they can with confidence be sited relatively close to centres of population where the demand for heat exists. Conceptual designs for such reactors have been developed in Sweden and France, and construction work has started on two reactors to supply district heating to the cities of Gorki and Vorenezh in the Soviet Union. There are said to be plans for the use of a considerable number of such plants in the Soviet Union but introduction in Sweden and France has been held up by a mixture of political and economic arguments.

The reactors for heat applications are light-water-cooled but rely on natural circulation in a large closed pool of water rather than circulation pumps in external loops, to remove the heat from the core region. In the Soviet design the coolant circulation is encouraged by some boiling as the water passes over the fuel and up a flume above the core – in effect this is using some of the energy of the reactor directly to circulate the water rather than converting it to electricity and using this to drive circulation pumps. Heat exchangers at the top of the pool transfer the bulk of the energy to a secondary circuit to produce hot water for district heating at between 120°C and 200°C, while the cooled primary water returns by natural circulation down the outside of the pool and back into the bottom of the reactor core.

With thermal power ratings in the range of 200 to 500 MW, these heat-only reactors represent a much smaller potential source of radioactivity than the large electricity producing power reactor where the thermal capacity in the core can be anything up to 3800 MW. At the same time it is possible to incorporate several layers of very substantial containment around the reactors because of their relatively small size. The large volume of water in the primary pool represents the first substantial layer of containment. Then there is usually some form of steel vessel inside a prestressed concrete outer vessel. Finally the plant can either be housed in a concrete containment building similar to those used for large power reactors or it can be partially or fully below ground level. It is protective measures such as these, together with the inherent safety characteristics of the reactors, which will allow them to be sited with confidence close enough to centres of population for district heating, and other low temperature industrial heat applications, to be competitive with alternative fossil fuels.

Ship propulsion reactors

Several hundreds of small pressurized water reactors in the power range up to about 50 MW, have been in use for many years for the propulsion of submarines and some naval surface vessels. Although detailed information is not available, all the indications are that they have operated extremely well. From the earliest days of civil nuclear development there has been interest in the possibility of using similar small reactors for commercial ships. An early demonstration ship, the *Savannah*, was operated by the Americans during the 1960s and a small ore carrier, the *Otto Hahn*, was operated by the Germans during the 1970s. A similar small demonstration ship, the *Mutsu*, was completed in Japan in 1976 but has been prevented from entering into operation due to persistent objections at potential harbour sites.

But the only nuclear-powered ships that have so far entered into routine commercial operation are three large icebreakers built by the Soviet Union. The first of these, the *Lenin*, demonstrated the technical feasibility of such a ship in the early 1960s using three small pressurized water reactors, but it was modernized in 1970 and is now operating with two reactors. Two more icebreakers, again each with two reactors delivering some 50 MWe to electric drives for the propellers, were put into service in 1974 and 1976 and another similar vessel is being built. In the course of trials one of these icebreakers has made a voyage to the North Pole and another has demonstrated the possibility of opening the trans-Arctic route to the Far East for all-year traffic. But now, in commercial operation, the main achievement of this small fleet of icebreakers has been to open up the north-west Arctic to all-year freight shipments in support of the development of rich natural gas reserves in the far north of the Soviet Union. They derive a special advantage from nuclear power because they can stay on station for long periods without a need to return to base for fuelling. Periods of thirteen months on station have been recorded by two of the icebreakers and it is estimated that, to do the same job, a diesel power icebreaker would have used 50 000 tonnes of oil and would have required ten interruptions for bunkering.

Icebreaking and propulsion of naval vessels such as submarines or aircraft carriers, are very special situations where features of nuclear power units make them particularly attractive – long endurance without the need for refuelling, no need for oxygen to support combustion and low noise levels. But for more traditional commercial shipping two factors detract from the economic case for nuclear power. The first of these is the significantly higher capital investment which is only likely to be recouped from lower fuel costs with large ships on long runs and with a relatively quick turnaround. The second factor is the need for expensive special-purpose port facilities to service the ships and, even though it may only be necessary to refuel reactors every three years, it would be necessary to operate a number of ships to justify the investment in such facilities.

Advanced Reactors

Nevertheless, with ever-increasing oil prices, the economic case for nuclear-powered ships, especially for high speed container ships or large oil tankers and ore carriers, is getting progressively better. Periodic economic studies, which attempt to take account of all the investment costs, usually quote the minimum size of ship on which a nuclear power unit would break even with more conventional propulsion. This breakeven point has been coming down steadily and is now, probably, for ships with displacements somewhere between 60 000 and 100 000 tonnes – well below the maximum size of many conventional ships now operating around the world. Factors that seem still to be inhibiting the commercial introduction of nuclear ships are, first of all, a sizable over-capacity of shipping which is discouraging the construction of anything, least of all a highly innovative ship and, secondly, the prospect of licensing delays similar to those encountered with land-based power stations which would be particularly disturbing for shipping companies looking for a reasonably quick return on investment.

The only firm plans at present are again in the Soviet Union and again for the special conditions of the partially ice-bound route down the Far Eastern coast. Work has started on an ice-strengthened ship of 60 000 tonnes displacement which can operate in ice up to 1 metre in thickness and will carry up to seventy-four lighters of 500-tonne displacement to ply to and fro between small ports and the mother ship. Since such a ship could alternatively carry 1330 standard 20-ft containers, its development will be watched with interest by operators on conventional shipping routes.

The pressurized water reactors used to power submarines use highly enriched uranium fuel in order to achieve a very compact core and a long period between refuellings. For commercial ships there is generally a little more space available and low enrichment of the fuel is usually more acceptable. In other respects ship reactors are essentially just small versions of land-based pressurized water reactors. The need for a compact containment for the plant and provision of a suitably inpenetrable crash protection around the containment, present challenging but not insurmountable problems to designers.

A number of innovative designs of ship reactors have been produced around the world. Usually the emphasis is on integrating more of the primary coolant circuit into the main reactor pressure vessel and avoiding as far as possible external loops of pipework connecting heat exchangers and circulation pumps. One such design has been developed and extensively tested as a land-based prototype in France. It uses a single steam generator incorporated in an enlarged lid of the pressure vessel above the reactor core. Circulation pumps are housed in short stubs on the side of the pressure vessel. Thus the whole of the primary circuit is effectively housed within a single pressure vessel. The first of a new class of submarine is now being equipped with this reactor.

Prospect for small land-based plants

There has long been an interest in, though as yet no market for, small land-based power reactors below the 600 to 1200 MWe range of commercial power stations used in industrial countries today. Smaller plants would be suitable for remote locations, as dedicated plants supplying steam and electricity to industrial complexes or for developing countries where the size of the electricity system is too small to incorporate the larger reactors which are now well-established in the industrial world. It is even envisaged that complete power stations might be built in barge-mounted form at shipyard facilities in industrial countries and taken by sea to coastal sites in developing countries.

Over the years, many conceptual designs have been put forward as contenders in this area. Obviously, land-based versions of ship reactors represent one possibility. One such example is a slightly larger version of the latest French submarine reactor with two steam generators connected by short double pipes to the pressure vessel and the circulation pumps housed in the bottom of the steam generator vessels. Another approach is to design a smaller and simpler version of the established commercial reactors. An example of this approach is a German design which would use boiling water reactor fuel assemblies of well-proven design in a down-rated core. Rather like the Soviet design of district heating reactor, boiling in the core region would be used to circulate the coolant by natural circulation through a single steam generator.

Another approach is to look at some of the early commercial power reactors which in their day were considered to be large units but by present day standards fall into the category of small reactors. Updated versions of such plants would offer a wealth of knowhow from the early plants, most of which are operating very well, having long since sorted out any teething problems. There are a number of single loop pressurized water reactors operating around the world with a capacity of around 450 MWe and the first generation pressurized water reactors still being built in considerable numbers in the Soviet Union and other Comecon countries are only 440 MWe. In Britain the first generation of gas-cooled Magnox reactors in concrete pressure vessels were built in pairs to provide stations with around 600 MWe total capacity but individual reactors of this type at 300 MWe offer another possibility that might have special attraction for a developing country due to the relatively low cost of natural uranium fuel. Similarly, the heavy water Candu reactors

Advanced Reactors

originally supplied to India by Canada but now being built independently, have a capacity of 200 MWe and, due to the modular nature of the pressure tube system, the Candu reactor can be offered in almost any size.

There is, therefore no shortage of possibilities for small nuclear power plants. The economic case for such plants is also becoming increasingly attractive since the only alternative in most developing countries is the use of very expensive oil, the import of which is already having a more devastating effect on their balance of payments than it does in the industrialized world. But even with a small nuclear power plant there remains a problem of a large initial capital cost and the need in a new country to establish an infrastructure of suitably trained people for operation and regulation of nuclear facilities.

The International Atomic Energy Agency (I.A.E.A.) has gone a long way in developing the sort of support that would be needed for the introduction of nuclear power in developing countries. In particular a comprehensive set of safety codes and standards has been developed and could be adopted as the basis for licensing plants in countries with no established background in the regulation and control of advanced engineering projects. At the same time the I.A.E.A. offers a wide range of technical support programmes for developing countries and these can certainly be extended to basic training of the people who would be needed to introduce nuclear power into a new country. The I.A.E.A. has also attempted to stimulate interest in the development of a standardized design which could be adopted by a number of countries. Such a development would help to reduce the capital cost by virtue of series production of components. But, in a highly competitive world, there is little likelihood of obtaining agreement among potential supplier countries on a single reactor design to be adopted as the standard. Rather it will need a number of developing countries to get together and agree to choose one of the competing designs as the standard that they will adopt for a series of plants.

Long-term prospects for fusion reactors

The most advanced type of power producing reactor on the horizon would be one making use of nuclear fusion rather than nuclear fission. Although this book is primarily concerned with nuclear energy derived from the fission of the large atoms, it is important to be aware of where we stand with the much greater potential source of energy that can be derived from the fusion of the smallest atoms. This is because fusion reactors could ultimately rival advanced fission reactors – notably the fast

Nuclear Energy

reactor – as a virtually unlimited supply of energy and it is necessary to keep in mind the timing and funding of the parallel development programmes.

The energy released when large atoms of uranium or plutonium are fissioned comes about because the stable structure of the two fragment atoms requires less force or binding energy to hold them together than the uranium atom from which they were formed. To maintain the balance between energy and matter, surplus binding energy is emitted during the fission. But at the other end of the scale of atomic weights we find an opposite situation in which the joining together or fusion of two hydrogen atoms produces a larger helium atom, which is held together with less binding energy than the sum of the two constituent parts and therefore, once again, the surplus energy is emitted.

The amount of energy released by such fusion is substantially greater than the energy released by fission but it is much more difficult to persuade two light atoms to stick together than it is to split a large atom. The only way this has been done to date is to use the force of a fission explosion to force sufficient hydrogen atoms together to create an even bigger explosion – in other words, the hydrogen bomb. Since the mid-1950s, however, there has been a great effort to try to demonstrate a controlled fusion reaction which could eventually be employed for useful power production.

Basically, the approach is to produce a very hot and dense concentration of the atoms to be fused, known as a *plasma*, in which the orbital electrons of the atoms have been stripped to leave positively-charged ionized atoms. A huge electrical discharge is used to produce the plasma, and powerful magnetic fields are used to hold the plasma together in a dense fireball. Since this is rather like trying to capture a lightning discharge inside a bottle, it is possible to realize why the development of controlled fusion reactors involves formidable technical problems. The various approaches that have been investigated over the past twenty-five years are too numerous to detail here, but the system that has emerged as the frontrunner in the attempt to produce a reactor in which more energy is derived from the fusion process than is consumed in generating and confining the plasma, is known as a Tokamak.

The Tokamak type of machine was first developed in the Soviet Union although it is a toroidal machine bearing a family resemblance to earlier machines developed in Britain and the United States. The plasma is produced as a closed ring inside a toroidal chamber by passing a large current through external coils, arranged as a massive transformer around the chamber. Further magnetic coils around the chamber are then brought into action to hold the ring of plasma in the centre of the toroidal chamber with an effective vacuum space between the plasma and the walls of the chamber. In this way there is no pathway for loss of energy from the plasma and it gets very hot. It can be visualized as a closed loop of wire sitting in space with a large electrical current going round and round to heat it up.

Other techniques are also available to help heat up the plasma, notably the injection into the plasma of high energy particles from an external source or radiofrequency heating. To initiate fusion of the atoms in the plasma on a scale that will produce more energy than is being used in the external magnetic coils and heating systems, it is necessary to hold the plasma in a stable dense state for around one second while it is heated up to temperatures of tens of millions of degrees – hotter than the temperature of the sun. To date this has not been achieved but there is good reason for confidence that it will be possible in large new Tokamak machines which will be starting to operate during the second half of the 1980s.

If one equates the demonstration of energy break-even in a fusion reactor to the first self-sustained fission chain reaction demonstrated by Enrico Fermi in 1942, it is possible to see why the production of useful power from fusion reactors is seen as a long-term development. There are, however, reasons for believing that once a fusion reactor has been demonstrated the period to development of a demonstration power-producing reactor could be less than it was in the case of fission. Not least of these reasons is the fund of advanced engineering experience obtained from the development of fission reactor power stations, much of which is applicable to the sort of problems that will be encountered in developing fusion reactor power stations. In addition, the fusion research teams around the world have already started work on conceptual designs of fusion power reactors with a view to identifying the areas where technical development will be needed (see Plate 6). Best estimates now put the likely timescale for demonstration of fusion energy break-even at between 1990 and 1995, the operation of power producing demonstration reactors around the turn of the century and, depending on economics relative to fast reactors, the beginning of commercial introduction between 2010 and 2025.

Fusion power reactors would initially use two isotopes of hydrogen – deuterium and tritium – as the fuel atoms to be fused together. Deuterium is, as we have seen in the case of heavy water production, recoverable from the small concentrations occurring in natural water, but tritium, which is an isotope of hydrogen with two neutrons and a proton in the nucleus, has to be produced by neutron bombardment of lithium atoms. One of the by-products of the fusion reaction is emission of high energy neutrons so it will be possible to breed tritium in a blanket of lithium situated around the reactor chamber of a fusion reactor rather as plutonium fuel is bred in the depleted uranium blanket around a fast reactor.

A pulsating plasma discharge in a Tokamak type of fusion reactor would emit surplus energy, partly in the form of direct radiation to the walls of the chamber but also as energy imparted to the neutrons flying off from fusing atoms. Absorption of both the direct radiation and the high energy neutrons in the walls of the toroidal chamber and in the lithium blanket around it would generate a lot of heat. This could then be extracted with coolant circuits, thence to raise steam and drive turbogenerators in much the same way as a fast reactor. The need for materials for the walls of a fusion reactor and the design of the blanket and cooling systems probably presents the largest technical challenge in the eventual development of fusion power because of the intensity of radiation and neutron bombardment that they will have to withstand.

One of the most attractive features of nuclear fusion compared with fission is the fact that it does not produce radioactive fission product wastes. This is not to say that fusion reactors will be free from problems of radioactive waste. The great intensity of neutron emission associated with the fusion reactor would be one or two orders of magnitude greater than the neutron emission from a fission reactor core of comparable power output. The structures around a fusion reactor will, therefore, become highly radioactive as a result of neutron irradiation. This will present a sizable problem for plant operators, especially as the lithium blanket has to be removed periodically for recovery of tritium but, for the most part, the radioactivity induced in steelwork and other materials around a fusion reactor will be relatively short lived and eventual disposal should present less of a problem than high level fission product wastes. Containment of the inventory of radioactive tritium that will accumulate in a fusion power plant could also represent something of a safety problem but here again it should be no more difficult than containment of radioactive materials in operating fission reactors.

Perhaps the biggest uncertainty about the eventual introduction of fusion power is the question of economics. All the present conceptual designs of fusion power reactors suggest that the capital cost will be very high, significantly higher than fast reactors which themselves are more capital-intensive than existing commercial nuclear power plants using thermal reactors. So much as the case for the introduction of fast reactors is based on compensation of the high capital charges by lower fuel costs, it will need yet lower fuel and operating costs for fusion reactors to break even with fast reactors.

Advanced Reactors

As far as ultimate energy resources are concerned, fusion can be described as having virtually infinite potential because of the practically unlimited quantities of deuterium in the oceans of the world. But this will depend on a second stage of development involving the fusion of two deuterium atoms rather than deuterium and tritium. While tritium is still being used as a fuel the ultimate resource limitation is the quantity of lithium available to breed tritium. In terms of the amount of energy they would produce, it is possible roughly to equate total world resources of lithium used in fusion reactors with uranium used in fast reactors. But since both these should meet global energy requirements for several thousand years, the question of resources is very academic.

8

Nuclear Industry Worldwide

The birth of any new industry, especially one based on high technology, is fraught with difficulties. The nuclear industry – which here we will take to include the complete spectrum of civil nuclear activities from research and development, through design, manufacture and erection of plants to operation of plants and supporting services – is no exception. Indeed, the nuclear industry has had to contend with many problems which are considerably larger in magnitude than the problems confronting other new industrial activities. Not least of these, in recent years, has been the development of hostility among groups of the public and a large section of the media and the excessive regulation which has been largely stimulated by this public opposition.

In the past many new industries have derived much of their basic technology from earlier military development. But in the case of the nuclear industry this link is nowhere near as significant as is usually represented. In the United States, Britain, Canada, France and the Soviet Union one can trace some link between military reactor development and early civil reactors, but there was a significant difference in the scale of civil development programmes. Where most other industries that have benefited from military spin-off have involved less sophisticated applications of the basic technology, the first civil nuclear plants were very much larger and usually more complicated than any of their military forebears. In areas as diverse as civil engineering, metallurgy and electronics, the early nuclear power programmes faced formidable new challenges and they have yielded a sizable technological spin-off to other industrial activities.

Partly because of the scale of nuclear power, but also for reasons of control and prestige, governments around the world have been very closely involved – some would argue too closely involved – in commercial development and exploitation. Since the basic principles of nuclear energy were made freely available at the first United Nations conference on the peaceful use of atomic energy in 1955, the pattern of development in most countries has involved the establishment of a national research and development agency. Impressive national research centres equipped with facilities which would have been beyond the means of private industry, were set up predominantly with direct government funding and, for the most part, direct employment of scientific staff. The exception, which may have influenced the smoothness of subsequent transfer of technical expertise to private industry, was the United States, where operation and staffing of national research centres relied to a greater extent on contracts with private industry.

The evolution of the sector of industry concerned with the design, manufacture and construction of nuclear power plants has been a fairly turbulent affair with many national complexities – some of which will be reviewed below. Generally the problem has been that the scale of the projects has been too big for all but the very largest companies and there has been a need to form industrial consortia of larger and larger size until, in most countries, there is now only one industrial group, often with a sizable government involvement.

At the operating end of the nuclear industry the countries with large nationalized electricity utilities clearly had the organizational structure to undertake large nuclear power projects. But in countries with a number of private utilities there has been some need for grouping of utilities to undertake nuclear projects and also a greater role for independent construction companies to provide project management.

The fuel cycle and other service industries are still to some extent in an evolutionary stage but here too the economy of scale is tending to dictate the formation of monopoly national, or even international, groupings. Since a large part of the expertise has been developed within national research and development organizations and also because it may involve sensitive technology, fuel cycle activities are, in many cases, undertaken by subsidiary companies formed as offshoots of the research and development organizations.

Britain

The first commercial nuclear power programme was launched in Britain in the mid-1950s following the successful start of operation of the Calder Hall power station. The four Magnox gas-cooled reactors at Calder Hall, and a later sister station at Chapel Cross, had been designed and built by a national research and development group which was to become the U.K.

● Operating (Jan 83)   35 reactors   total capacity 13.5 GWe
○ Under construction or planned   8 reactors   total capacity 5.5 GWe

Figure 8.1 Nuclear power plants in Britain

Atomic Energy Authority (A.E.A.), for the dual role of military production of plutonium and demonstration of electricity production on a commercial scale. To meet the first series of orders for purely civil nuclear power stations from the nationalized utilities, the Central Electricity Generating Board (C.E.G.B.) and the South of Scotland Electricity Board (S.S.E.B.), there was a need to form private industrial design and construction groups, known as consortia, because no one industrial group had the necessary range of expertise nor the size to undertake such large and technically advanced projects. No less than five industrial consortia were set up and each developed different designs of power station from the basic Calder Hall concept. Although these early commercial nuclear power stations suffered some delays and increases in cost during construction, they all eventually started to perform very well and are today producing the country's cheapest electricity.

Nuclear Industry Worldwide

The evolution of the British nuclear industry from these early beginnings and the organizational problems encountered are illustrative of difficulties encountered in many industrialized countries during a period of very rapid development. Economy of scale became a dominant economic driving force and unit sizes in all types of industrial plant saw dramatic increases. The first two commercial nuclear power plants in Britain put into service at Berkeley and Bradwell were, in their day, considered to be large power stations. They both had a capacity of less than 300 MWe supplied from two reactor units and, respectively, four and six turbogenerator units. Today there are several nuclear power stations operating in other countries with single reactor and turbogenerator units rated at 1300 MWe.

One obvious consequence of this rapid escalation in unit size of both nuclear reactors and turbogenerators was that there were less orders and fewer prospects for competing industrial groups. There were therefore mergers of the industrial consortia, accompanied in most cases by parallel mergers of competing member companies in the groups. For a while in the 1960s there was an effort to maintain two competing consortia and some thought was also given to establishing international links which might have resulted in two competing industrial groups in the European marketplace. But after much wrangling, the final outcome in Britain has been the establishment of a single National Nuclear Corporation (N.N.C.) in which the main construction and manufacturing companies have shareholdings. The state also has an interest in the form of 35 per cent of N.N.C. shares held by the A.E.A.

Another major factor in the development of the British nuclear industry has been the dominant position of the C.E.G.B. as major customer for nuclear power plants. The C.E.G.B. has become one of the world's largest operators of nuclear power plants and, in the process, has developed a very strong in-house engineering capability for supervision of design and construction as well as operation. On the latest nuclear power plant projects the C.E.G.B. is assuming the overall management responsibility while the N.N.C. is acting merely as agent for detailed design of the nuclear sections of the power stations and for management of the associated subcontracts.

One of the most agonizing problems that has bedevilled the development of the nuclear industry in Britain has been repeated arguments about the choice of reactor to be used. The original choice of gas-cooled reactors was largely dictated by the need to use natural uranium fuel, but when enriched uranium became commercially available there was a choice between development of more advanced gas-cooled reactors or light water reactors of the type developed in the United States. There was also interest in the promising development of heavy water reactors in Canada. Unfortunately the British industry has been, and still is to some extent, split into several different factions, each arguing the undoubted merits of different reactor options but unable to achieve a sufficient body of support to make a real success of any one.

Arguably, in such a situation, market forces and in particular the preferred economic choice of the utility that has to operate the plant, should be allowed to decide. But it is difficult to establish a clear economic case without building and operating a plant under prevailing British conditions. No one faction within the industry has proved strong enough to get its preferred reactor adopted and so there has been a fallback on government to make the choices, with all the conflicting political pressures that that involves.

One factor that fuelled the continuing controversy over reactor choice in Britain was the rather unhappy experience with construction of the first commercial advanced gas-cooled reactors. Although these plants are now starting to operate well and economically, they suffered many technical problems and delays during construction – especially the first plant at Dungeness B which was more than ten years behind its original, rather optimistic, schedule. The problems led to a considerable loss of confidence in the A.G.R. and in the still more advanced high temperature gas-cooled reactor which might have followed.

Another factor that has delayed a clear choice of reactor has been a lack of urgency to order new electrical generating capacity of any kind. A sharp decline in the rate of growth of demand for electricity was caused initially by the discovery of natural gas in the North Sea and its rapid penetration of the domestic energy market, and then by a slowdown in economic growth rate which got progressively worse after the 1973 oil crisis. This has produced an apparent surplus of electrical generating capacity and, although some of this is from oil-fired plants committed during the 1960s when it was thought that low oil prices would continue, it has been possible for politicians to put off decisions on new nuclear plants and for the arguments to continue.

The nuclear power plant construction programme presently envisaged in Britain for the 1980s will include the completion of two new A.G.R. stations – each with twin 660 MWe reactors – which were committed in 1979 and are due to operate in 1987. A government decision at the end of 1979 decided that, subject to the necessary licensing approvals, the next order should be for a pressurized water reactor, based on a licence from the United States. The most important of the necessary approvals will be decided by the outcome of a public inquiry which started in January 1983 to consider the siting of such a plant at Sizewell B. Beyond this, official projections of energy demand range widely – the lowest derived from an economic growth rate of ½ per cent per year would still call for five more units to be ordered in the next ten years while a growth rate of 2½ per cent per year could produce a need for as many as twenty units in that period. The most likely forecast, however, would require the ordering of about 9 GWe of capacity between 1983 and 1993. The actual placing of plant orders will, of course, be dictated by continually-revised forward estimates of electricity demand.

In the prevailing period of recession and abnormally low industrial production in the country, even these modest growth rates look to be on the high side and the outline ordering programme for new plants may well slip back. If, however, longer-term provision is to be made for future economic recovery, then a nuclear power programme of the scale envisaged is considered to be about the minimum level needed to maintain a viable domestic nuclear industry.

The development of the nuclear fuel cycle industry in Britain has been a somewhat happier story. While lack of clear decisions on reactors and the rate of ordering of new nuclear power plants can present difficult problems for the timing of new investment in fuel cycle facilities, at least there is a predictable on-going demand for fuel for operating nuclear power plants. As a result of the early start in commercial exploitation of nuclear energy a very strong technological base in all the fuel cycle activities was established by the Atomic Energy Authority and this basic technology can be directed towards the production of fuel for virtually any type of reactor. These activities were put on to a more commercial footing in 1971 when British Nuclear Fuels Limited (B.N.F.L.) was set up to take over the former activities of the A.E.A. Production Group. Provision for private participation was made in the act establishing B.N.F.L., but the company has remained a wholly state-owned organization with the shares held by the Department of Energy.

Since its formation B.N.F.L. has operated with respectable profitability and has also embarked on a large investment programme in new and refurbished plants. The largest development is the building of a new reprocessing plant at Windscale (now renamed Sellafield) to deal with oxide spent fuel from the British A.G.R.s and also light water reactor fuel for foreign customers. Associated with this development will be large new ponds for temporary storage of spent fuel prior to reprocessing and a new plant for vitrification of the high level waste products separated during reprocessing. At Capenhurst B.N.F.L. is the operating company involved with the installation and operation of a new centrifuge enrichment plant which forms part of an international programme in collaboration with Dutch and German partners. The international company,

Urenco, which is responsible for marketing these enrichment services, also has its headquarters in Britain. Finally,, B.N.F.L. operates a very large complex at Springfields which includes plant for refining uranium ores, for conversion of uranium to uranium hexafluoride and for the fabrication of nuclear fuel assemblies for just about any kind of commercial nuclear reactor.

The economic case for nuclear power in Britain was initially rather marginal and the early programme was influenced by worries about oil supplies stimulated by the Suez crisis. By the time these plants were put into service worries about oil supplies had disappeared and it looked as if they would have difficulty competing with what was then low priced oil. At the same time, the rising cost of coal production ensured that the main source of the country's electricity production became progressively more expensive than electricity from nuclear plants. In spite of heavy costs incurred by technical problems and delays during construction, the second nuclear power programme of A.G.R. construction is maintaining the trend in generating cost advantage relative to coal. In the meantime we have seen gigantic increases in the price of oil which is now far more expensive than either coal or nuclear power. Projections for future plants suggest that the advantage of nuclear power over domestic coal will continue to increase while nobody seriously contemplates the building of any new oil-fired power stations because of the excessive fuel costs.

A feature of the economics of nuclear power which is different to fossil-fuelled power stations is that their relative generating cost advantage appears to improve as they get older. In a period of rapid inflation, this is a characteristic of plants with a high initial capital cost but low fuel costs. This will become even more marked when, as is likely, the operation of the plants is extended beyond the period over which capital charges have been spread. In the case of the early British plants this book life is only twenty years and, even if some periods of shut-down are called for to make the repairs and modifications needed to maintain safe operation, their economic advantage looks ever more attractive. This feature has led to criticism of the method of arriving at generating costs and it has been suggested that some form of accounting which makes allowance for inflation of the eventual plant replacement cost should be used. The latest current cost accounting methods of the C.E.G.B. go part of the way towards meeting this criticism and the figures still indicate a relative and increasing advantage for nuclear power over coal. The comparison might look even more favourable if all the government money spent in the past in development of coal mines and coal subsidies as well as the likely cost of future replacement of old coal mines, were to be factored into the cost of electricity production from coal.

France

The second country to launch a programme of commercial nuclear power station construction was France. The pattern of development was at first similar to Britain, with the adoption of a gas-cooled reactor fuelled with natural uranium. The system was developed by a national research and development agency, the Commissariat à l'Energie Atomique (C.E.A.), initially for military production of plutonium, and was then built for the state utility, Electricité de France (E.d.F.), by a number of different industrial consortia. Although these early stations have performed well, they ran into stiff competition in the mid-1960s from oil-fired power stations for which, at that time, there was thought to be a plentiful supply of low cost fuel available from North Africa and the Middle East. As a result there was a period of argument, quite as intense as that which has persisted in Britain, about the choice of reactor system. In the case of France, however, these arguments only lasted for a few

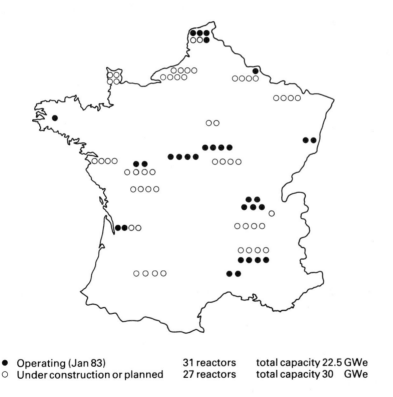

• Operating (Jan 83)   31 reactors   total capacity 22.5 GWe
○ Under construction or planned 27 reactors   total capacity 30   GWe

Figure 8.2 Nuclear power plants in France

years and were eventually resolved by the customer, E.d.F., taking a firm stand on its wish to adopt light water reactors.

A period of fairly painful readjustment by the industrial groups followed. Initially two competing groups were formed to offer pressurized and boiling water reactors based on Westinghouse and General Electric licences from the United States but, largely as a result of government intervention, there was a further readjustment involving a complex series of forced mergers, reallocation of shares and redefinition of areas of activity. A company called Framatome, which had originally been set up in partnership by French companies and Westinghouse to build an early P.W.R. power station as a joint Franco-Belgian project at Chooz, became the sole supplier of P.W.R. steam supply systems for the French commercial power stations. The large heavy engineering group, Creusot Loire, has a majority shareholding in the reorganized company and over the years the minority Westinghouse holding has been taken over by French interests, notably the C.E.A. which now holds 35 per cent of the shares on behalf of the state. Due, apparently, to this existing state holding, Framatome was not included in the nationalization programme when President François Mitterrand came to power in 1981.

The programme of nuclear power plant construction which has been carried out in France through the 1970s, and which is continued through the 1980s, has become a model which is the envy of the nuclear industry in the rest of the world. True the French industry had a lot going for it. The unification of the industry, although painful, did produce a single-mindedness of purpose and a powerful technical effort in support of the programme. There was strong support from President Valéry Giscard d'Estaing, and the highly centralized government of the country was able to make clear decisions without time-consuming prevarication. There was a growing demand for electricity and, due to the lack of any significant indigenous energy resources, there was a clear recognition of nuclear energy as the only available means of combatting the country's heavy dependence on imported oil. E.d.F. was therefore able to offer the industry a large programme of series ordering of highly standardized plants, and Framatome was able to respond by investing in impressive purpose-built facilities for the manufacture of major components such as pressure vessels and steam generators. Another important factor in the success of this programme has been firm management by E.d.F. and a clear definition of responsibilities at the construction sites.

By 1980 new nuclear plants were being commissioned in France at a rate of one every two months. Construction periods were down to five-and-a-half years and the time taken to get new plants from initial start-up to steady supply of electricity at their full power rating was down to a couple of months. As a result the contribution of nuclear energy to electricity production in France has increased dramatically from under 10 per cent supplied by the first generation gas-cooled reactors in the mid-1970s to a total of over 40 per cent in 1982. This nuclear contribution will further increase to 50 per cent in 1985 and 75 per cent by the end of the decade. More importantly, by the end of the decade nuclear energy will be supplying nearly one third of France's total energy needs, with dependence on oil reduced from an alarming level of 80 per cent in 1973 to around 30 per cent.

The importance of the nuclear contribution to the French economy was clearly recognized when the new Socialist government reviewed the country's energy policy in 1981. In spite of having expressed reservations about the size of the nuclear power programme in previous political declarations, the Socialists initially only cut the rate of ordering of new nuclear plants by about one third – a reduction which might well have occurred anyway as a result of revised projections of growth in energy consumption during the 1980s. However, further slowdown in economic growth rate suggests that continued ordering at the rate of only one 1300 MWe plant per year may be the most that can be justified during the second half of the 1980s and then only on the basis of maintaining support for an industry that is vital to the long term interests of the country.

The success of the latest French nuclear power plant construction programme owes a great deal to the adoption of a well proven reactor design and replication of a large number of virtually identical units. There are now thirty-four reactors with a capacity of 900 MWe operating or under construction in France and a further eighteen with a capacity of 1300 MWe are under construction or planned. All these reactors are being built under licence from the Westinghouse company for P.W.R. designs with three and four primary coolant loops. But in the process of carrying out such a large programme, Framatome has built up its own technical expertise and experience to the point where it is introducing evolutionary design improvements. A new 1300 MWe design of P.W.R., which is claimed to be entirely French, has been developed and is likely to be installed during the 1980s. The licensing agreement with Westinghouse, which was due to expire in 1982, has been renegotiated and replaced with a two way technical collaboration arrangement. In parallel with the design and construction work of Framatome, the C.E.A. has been providing important technical support and has built impressive test facilities at the Cadarache research centre for qualification and development of P.W.R. components and for exhaustive safety studies.

On the strength of its large domestic programme the French industry has been able to establish a strong position in the world market for nuclear power plants. Although the market in countries without their own nuclear industry has been relatively small in recent years, Framatome has managed to obtain orders for eight units – although construction of two of these in Iran was subsequently frozen – and is actively bidding in important new markets such as China and Egypt. Recognizing an inevitable slackening in the domestic demand for new plants by the end of the 1980s, the hope is that these new markets will develop to take up some of the available capacity in the French manufacturing facilities.

With an early start in nuclear power and a large programme France has naturally developed a complete range of skills in the nuclear fuel cycle industry. Most of this development work was carried out by the C.E.A. but there has also been active private industrial participation, notably by a large chemical group, Pechiney Ugine Kuhlmann (P.U.K.), which through a number of subsidiaries has been involved in both uranium mining and fuel fabrication. The fuel cycle activities of the C.E.A. were split off into a wholly-owned subsidiary called Cogema (Compagnie Generale des Matières Nucleaires) in 1969. Following nationalization of the P.U.K. group in 1982, further rationalization of French fuel cycle activities under the Cogema umbrella is expected.

A number of deposits of uranium have been found in France and, although small in comparison with the main uranium producing regions of the world, they have been developed commercially and have provided the country with considerable expertise in exploration, mining and processing techniques. This has been put to good use through joint ventures in various parts of the world and in particular the former French colonies in central Africa. At the beginning of the 1980s about half of the uranium requirements for the country's nuclear power programme could be supplied from domestic resources and the rest came from central Africa. But with the rapid expansion of the nuclear power programme, more and more will be coming during the 1980s from the larger and cheaper sources of uranium in countries such as Australia, Canada and South Africa.

Uranium enrichment by the gaseous diffusion process was developed by the C.E.A. initially for military purposes. But about the time that the British, Dutch and Germans were forming Urenco as a European company to exploit centrifuge enrichment on a commercial basis, the French formed a rival international company, Eurodif, with Italian, Belgian and Spanish participation to build a large commercial enrichment plant using their gaseous diffusion technology. The differences between the rival processes necessitates commitment to a very large-scale project in the case of gaseous diffusion while centrifuge capacity can be built up progressively in response to market demand. The Eurodif plant, located at Tricastin, therefore represents a very bold development. Operation started in 1969 and by 1982 the project was completed, giving a potential capacity of 10.8 million separative work units per year. In the event nuclear power programmes in the countries of the other partners have not developed as quickly as expected and the Eurodif plant is likely to be operating beyond the output immediately required during most of the 1980s.

The other disadvantage of gaseous diffusion relative to centrifuge enrichment is the high consumption of electricity – about ten times higher for a given output – but the French have overcome the cost penalty that this might have incurred by the simple expedient of building a large nuclear power station with four 900 MWe alongside the enrichment plant to supply cheap electricity. Since the enrichment process is in effect increasing the energy content of uranium fuel, this combination can be considered as a method of storing the energy from the adjoining nuclear power plants for later release in a much larger number of plants.

The French industry has developed a great deal of skill in fabrication of fuel assemblies for all types of nuclear reactors. Continuing commercial production of natural uranium fuel for the early gas-cooled reactors is now concentrated at one plant at Annecy. For the introduction of the large programme of P.W.R.s a complex shareholding arrangement provided a controlling French interest in a group with a Westinghouse licence and fabrication facilities at Mol in Belgium and Roman in southern France. More recently, a joint company called Fragema has been formed by Framatome and Cogema to produce a wholly French design of fuel assembly for P.W.R.s in a new fabrication plant. It is probable, however, that the complex interrelationships between these fuel fabrication organizations will eventually be rationalized into a single group and that there will be a phased transition from a Westinghouse to French design of fuel.

The C.E.A. set up two reprocessing plants for natural uranium spent fuel, one at Marcoule in the south and the other at Cap La Hague on the Channel coast. These are now managed by Cogema and the plant at La Hague is being developed with a view to handling all the reprocessing of oxide fuel from P.W.R. plants while reprocessing of natural uranium is concentrated at Marcoule. Fuel from other countries is already being reprocessed at La Hague and the development of the plant will provide additional capacity to fulfil a substantial volume of reprocessing business that has been contracted from foreign customers – notably Japanese and German.

Nuclear Energy

The overriding case for nuclear power in France has been the need to find a source of energy which will give this very independently-minded country a greater degree of self-sufficiency. This is a particularly acute problem for an advanced industrial country that is trying to maintain a high standard of living with remarkably few indigenous sources of energy. France has a small amount of expensive coal, more or less fully exploited hydro-resources in the Alps, and no domestic oil or gas nor any immediate prospect of finding any in France's offshore waters. Given therefore that the bulk of energy resources must be imported, uranium is the cheapest and most freely available on the world market. And in the long term the stockpiles of depleted uranium already accumulated in the country, together with the modest domestic reserves of uranium, represent a virtually unlimited source of energy if used to feed fast reactor power plants.

As a result of these strategic considerations, the French nuclear power programme is designed not just to meet a growth in electricity demand related to the country's economic growth, but to attract an increasing number of end users of energy away from oil and gas to electricity. Perhaps the greatest success of the current programme is that by completing a large number of standard plants to time and cost and achieving excellent operating performances quickly, the cost advantage of nuclear electricity generation is starting to have a significant impact on the overall cost of electricity on the E.d.F. system and the economic incentive to switch to electricity wherever possible has been clearly established. Figures issued by E.d.F. each year now show a steadily increasing cost advantage for nuclear electricity. By 1981 the cost of nuclear electricity was between a half and one third that from coal and less than one third of the cost of electricity from oil. But more significantly, the cost of electricity to consumers in France was the lowest in Western Europe and it was the only country showing a growth in electricity consumption – 3.8 per cent in 1981 – in spite of the general stagnation of economic growth. Also in 1981, France reversed the situation of the previous four years when they had been a net importer of electricity from neighbouring countries, becoming a net exporter – a trend which is likely to increase during the 1980s.

This success story has not been without its moments of concern. The first of the new P.W.R. plants were about eighteen months late in coming into operation and created a very critical period for E.d.F., due partly to the narrow margin of reserve generating capacity but also to the huge financial burden of a large and very capital-intensive construction programme. There was also a period of concern when a problem of very small cracks was discovered in pressure vessels and threatened to delay start-up of several plants. But, after a stormy public debate, it was established that the cracks represented no safety problem. Now, although the financial investment burden being carried by E.d.F. is still very substantial, the operating plants are yielding a handsome payback in low cost electricity. The main economic challenge in the coming years will be to maintain the clear cost advantage of the nuclear plants even when they are following the daily and seasonal variations in electricity demand rather than operating continuously at full power to meet the country's base load requirements.

United States

An abundance of low cost energy resources provided the United States with very low cost electricity in the 1950s and there was no great incentive for the many independently operated electricity utilities to switch to nuclear power. The long-term potential was, however, recognized both by the federal government and by the more far-sighted utilities. The government-funded Atomic Energy Commission (A.E.C.) therefore

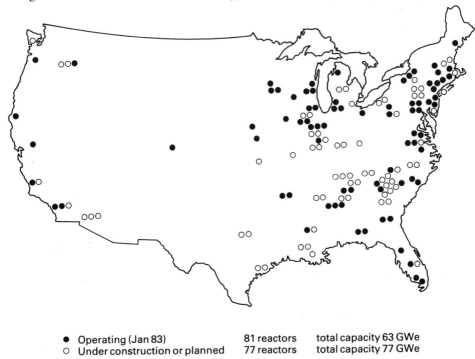

• Operating (Jan 83)   81 reactors   total capacity 63 GWe
○ Under construction or planned   77 reactors   total capacity 77 GWe

Figure 8.3 Nuclear power plants in America

undertook a very broadly-based programme of technological development at a number of large national research centres and also helped a number of private utilities with the funding of early nuclear power demonstration projects.

It is not often appreciated that a very wide variety of different types of reactor was investigated at this time as potential contenders for commercial power production but, perhaps significantly, the final choices of which systems should be developed were taken by large commercially-minded companies with a long tradition of meeting the needs of the electric utilities. These companies were certainly helped in making their choice by participation as major contractors on many of the different development and demonstration projects funded by the A.E.C. This was probably a more important source of technical know-how than the submarine programme, often attributed as the forebear of the American civil reactors, because the companies involved were severely restricted for security reasons from making direct civil use of the submarine reactor technology.

The number of reactor options available to the American nuclear industry was also extended considerably by the availability of enriched uranium. Three large gaseous diffusion plants which were built initially for military purposes were also able to supply low enrichment material for civil purposes. Freed from development and investment charges and with plentiful supplies of low cost electricity from coal and hydro sources, these plants were able to offer enrichment services at prices which had a strong influence on the choice of reactor system.

Two major electrical engineering companies, General Electric (G.E.) and Westinghouse, took the lead in commercial development of, respectively, boiling water and pressurized water reactors. Two other large companies in the heavy mechanical engineering field, Babcock & Wilcox and Combustion Engineering, with well-established and continuing business in the supply of conventional boilers to utilities, established their own capability to supply P.W.R.s. These two companies were also able to apply their heavy engineering experience to the manufacture of major components, notably reactor pressure vessels, for the plants of G.E. and Westinghouse as well as their own plants.

Based on the knowledge developed from the early demonstration projects built for American utilities, and with the help of government fuel leasing under the Atoms for Peace programme, G.E. and Westinghouse also had some success selling similar relatively small demonstration reactors to other countries interested in the peaceful use of nuclear energy. The countries included Italy, Japan, West Germany, Belgium, Spain and the Netherlands and the generally very good operating performance of these early plants contributed a great deal to the establishment of a fund of technical know-how which was of benefit to development of commercial nuclear power both in the United States and the other participating countries.

But it was not until the early 1960s that the true breakthrough into the commercial power plant market came, with sales to utilities unsupported by the government demonstration programme and in competition with conventional coal- and oil-fired power stations. It was also necessary for G.E. and Westinghouse to offer the first commercial power stations on fixed-price turnkey contracts under which they would assume overall responsibility for complete projects. This contrasted with the more traditional non-turnkey contracts for conventional power stations which involved architect/engineering companies, or the utilities themselves, assuming the overall project management and issuing subcontracts to boiler makers and turbogenerator manufacturers as well as a multitude of smaller subcontracts. The two companies were later to admit that they had lost heavily on the first few turnkey contracts for nuclear power stations and it was not until they reverted to non-turkey contracting that they started to recoup some of their heavy investments.

But they did open a floodgate to utility ordering of nuclear power plants throughout the 1960s. It was a period of very rapid growth in demand for electricity and, apart from establishing a competitive advantage in most parts of the country, nuclear power overcame growing concern about environmental pollution in the form of sulphur dioxide and other emissions from conventional power stations. The pattern of ordering was not unlike that of the civil aircraft business, with utilities scrambling to get in line with firm orders and options for follow-on units. By the mid-1970s, however, the worldwide recession caused by the oil crisis had resulted in a dramatic slowdown in growth of demand for electricity, and environmental pressures were being directed against nuclear power. Despite a greatly increased competitive advantage brought about by oil price rises, many of the nuclear orders and options from the late 1960s were postponed or cancelled altogether.

In recent years American utilities have been confronted by a unique financial problem which has been preventing them from ordering new nuclear plants. Although they operate for the most part as independent investor-owned companies they are heavily regulated by federal and state agencies. These dictate the tariffs that can be charged for electricity and also, in most cases, prohibit the utilities from passing on any of the capital cost charges on investment in new plant until that plant starts feeding electricity into the system. When utility bonds were highly favoured by small investors and new plants could be brought into operation in five and a half to six years after the initial decision to build, this did not present too much of a problem. But the cumbersome nuclear licensing processes now mean that it takes anything from ten to twelve years from commitment to operation of a new plant; the severely regulated tariffs have hit the cash flow of the utilities and put them in a position where they cannot compete with the high interest rates being payed by other types of investment; and the very high interest rates on borrowings from the money markets impose an impossible burden on the utilities during the construction of plants. So, while most utilities are still convinced that nuclear plants will yield a handsome pay-back on investment over their operating life, they are encountering considerable difficulty in finding the money to cover the construction period.

Despite suffering from ups and downs in the ordering of nuclear plants, caused by a complex mixture of over-regulation and free market forces strongly influenced by changing political climates and public opinion, the American nuclear industry has had a very considerable success. It still has a sizable volume of business in the pipeline which will keep it going until the late 1980s, even if there is no immediate change in the current recession. By the beginning of the 1980s there were some seventy-eight reactors in operation with a total capacity of just under 60 000 MWe, supplying between 11 and 12 per cent of the country's electricity requirements. Projects already under construction or ordered and awaiting licensing authorization will nearly treble the generating capacity during this decade when nuclear power will meet between 30 and 40 per cent of electricity demand. Bearing in mind that the United States consumes almost as much electricity as the rest of the Western world put together, the achievement of the nuclear industry in a relatively short time seems very considerable.

Arguably, it could have been larger – projections in the 1960s foresaw an installed capacity in 1980 of almost double the level actually achieved. While this reduction in the actual nuclear contribution corresponds roughly to a comparable scaling down of growth in electricity consumption, the United States would certainly have benefited from the additional nuclear capacity, even at the expense of making a number of oil- and gas-fired power stations redundant. The missing nuclear capacity would have replaced oil consumption almost equal to the country's level of oil imports.

The pattern of use of nuclear power across the United States varies markedly from state to state. The east coast and mid-west states have shown considerable interest mainly due to lack of indigenous energy resources, and an early recognition of the value of the nuclear option in maintaining clean air conditions. Some of the smaller New England states generate from 30 to over 70 per cent of their electricity from nuclear energy and consumers are beginning to see the benefits in relatively low electricity bills as other states suffer from the cost of oil dependence. Of the larger states, several of which have levels of energy consumption which are comparable with a medium-sized industrial country, Illinois has made most progress with nearly 30 per cent of electricity produced from nuclear energy and dependence on oil and gas cut to around 8 per cent. New York state gets nearly 20 per cent of electricity from nuclear energy but still has a heavy dependence of around 40 per cent on oil and gas. Pennsylvania and Ohio produce 80 to 90 per cent of their electricity from coal, with the balance fairly evenly spread among other sources including nuclear energy. California, the state with the largest consumption of energy, was prevented from adopting nuclear generation largely by highly restrictive state legislation, and gets a mere 3 per cent from nuclear power while oil and gas still account for 60 per cent. The balance comes mainly from hydropower. It may be significant that Californians pay around 30 per cent more for their electricity than the national average.

There is no doubt that the Three Mile Island (T.M.I.) accident in 1979 had a large impact on the development of the American nuclear industry. It happened at a time when the administration of President Carter, which had rated nuclear power as the energy option of last resort, was just coming round to the fact that they would have to accept this last resort in the economic interests of the country. A Presidential statement on energy policy had been expected to sweep away some of the administrative delays in the licensing of nuclear plants that had largely resulted from the efforts of the regulators to be seen to be responding to the much-publicized concerns of the opponents of nuclear power. T.M.I. prevented any early move in that direction and started a soul-searching review of all aspects of nuclear safety.

American industry in general is as likely to make mistakes as the industrial sector in other countries, but when it does it seems to be very thorough in its analysis of the lessons to be learnt from the mistakes. The T.M.I. accident resulted in very severe damage to the nuclear reactor but, contrary to popular belief, was a very long way from causing any direct harm to people either working on the plant or living nearby. The biggest damage has been the huge cost that is likely to be incurred in cleaning up and repairing the plant and the even larger cost of replacing the electricity that it would have generated from more expensive sources. A substantial bonus is a very complete review of all nuclear plant design, construction and operation which has been communicated to the whole world. As a result, all nuclear plants are probably even safer than they already were.

For the American nuclear industry the main lessons from T.M.I. have been learnt by the utilities operating nuclear power plants. Many of these utilities had assumed that once they had satisfied all the demanding criteria of the Nuclear Regulatory Commission and once their operators had passed the necessary examinations, they had a safe plant. T.M.I. has resulted in a significant change of attitude which recognizes: the need to have a fuller comprehension of plant safety rather than a checklist of satisfied criteria; the need for operators to have a fuller understanding of what is going on inside their plant both from better technical training and from improved instrumentation; the need for improved management especially over maintenance operations; and the need for better communication between utilities about small operational incidents which might be precursors of more serious accidents. One important outcome has been the establishment of an Institute of Nuclear Power Operations which is conducting regular critical reviews of all the nuclear power plants in the country. It is also operating a system for rapid communication and analysis of operating incidents which includes links with operators in several other countries.

The process of taking on board the lessons learnt from T.M.I. has been painstaking and time-consuming especially within the Nuclear Regulatory Commission. But, under the more favourable climate of the Reagan administration, there are signs of a slow revival which may clear the backlog of licensing of projects in the pipeline by the mid-1980s and might lead to new orders for nuclear plants starting to flow in the second half of the 1980s.

The nuclear fuel cycle industry in America has suffered to some extent from the efforts of successive administrations to transfer as much as possible to the private sector and to exercise control over sensitive technology. The private industrial sector needed little encouragement to get into the uranium mining and exploration business in the late 1940s and early 1950s in response to a large military demand. Most of the established mining and oil companies, as well as many smaller newcomers, enjoyed relative ease in discovering substantial reserves – notably in the states of Wyoming, Colorado and New Mexico. Even with more recent large discoveries in Canada, Australia and the African continent, the United States still has the largest proven reserves of low cost uranium and the largest rate of production.

But the American uranium supply industry went through a severe slump in the late 1950s when the military demand came to an end and before the civil market had built up to a significant level. To protect the domestic industry, and in particular to encourage continued exploration against the day when the civil demand would outgrow the early military requirements, the government stepped in to impose an embargo on imported uranium. Only in recent years has this import embargo been phased out and in the meantime it precipitated a legal battle of historic proportions between American interests and uranium producers in the rest of the world who had clubbed together to protect a minimum price level during the slump years. The problem arose when the market price of uranium rose rapidly after the 1973 oil crisis. The Westinghouse company, which had concluded long-term fuel supply contracts with many of its nuclear power plant sales, had not fully covered all these contracts with firm uranium supplies. Faced with the prospect of huge financial losses the company attempted to revoke its fuel supply contracts and charged foreign uranium producers with the formation of a cartel. The resulting legal battles were only resolved finally in 1981 with a number of out-of-court settlements, ironically at a time when the uranium market and price were again declining due to the slippage of nuclear power plant construction programmes around the world.

Although the American government looked very seriously at the possibility of transferring the uranium enrichment business to the private sector in the late 1960s, it has in the event retained ownership of three large enrichment plants. These were built in the first place for military production of high enrichment material but have been progressively upgraded for large volume production of low enrichment material. Based on this large available capacity the government established the pattern for long-term commercial contracts for enrichment services, under which operators of nuclear power plants supply uranium feed and pay a charge to have it enriched to the appropriate level before it is transferred to commercial fuel fabricators. These contracts, now administered by the Department of Energy, have dominated the world market during the 1970s with only a small portion of the requirements in the non-Communist world being met from contracts with the Soviet Union. During the 1980s, competition from the large Eurodif plant in France and the smaller centrifuge plants of Urenco in Britain and Holland, will be growing but the American plants are still likely to be accounting for 60 to 70 per cent of the Western market at the end of the decade.

Fabrication of fuel assemblies has from the outset been undertaken by private industry, primarily the four main reactor manufacturers who have included initial loadings of fuel in their nuclear power plant supply contracts and the associated warranties on performance. But the American market has also seen the establishment, by subsidiaries of the major oil companies, of fuel fabricators competing to supply the replacement fuel for subsequent loadings on a number of power plants. This is an area where there is scope for worthwhile improvements in performance of the fuel with evolutionary changes in design and, even though the market has not yet grown to the size that had once been projected, it is a continuing market which is likely to attract lively competition.

The sector of the nuclear fuel cycle industry that has encountered most problems in the United States is reprocessing. At the outset it was seen as an activity for the private sector and a number of companies showed interest in setting up commercial plants. A small demonstration plant was operated on a commercial basis for a number of years but when it was shut down for upgrading the political climate and licensing requirements had changed so drastically that the prospects of reactivating the plant were finally abandoned. Another plant employing novel technology was built but encountered technical problems at the commissioning stage and was also abandoned as a commercial reprocessing facility.

The first truly commercial plant, with a capacity of 1400 tonnes per year, was virtually completed at Barnwell, South Carolina, and was awaiting clear decisions from the government on the policy for commercial recycling of plutonium before installation of the final sections of the plant. But before any such policy could be formulated, President Carter came to power and in 1977 renounced commercial reprocessing and plutonium recycling. President Reagan has now reversed this decision but at the same time has indicated that commercial reprocessing should be undertaken by private industry without government support. Not surprisingly, the private industrial sector is no longer very enthusiastic about the now substantial cost of completing the Barnwell plant unless there is some sort of government involvement to cover the future uncertainties of political changes.

In the absence of commercial reprocessing and clear policy decisions on plutonium recycling, there has been a substantial build-up of spent fuel at the nuclear power plants operating in the United States. The introduction of special racks, which allow storage of spent fuel from ten or more years of plant operation in ponds at the power stations, means that the build-up does not yet present severe problems. But it is recognized that there will be a need for interim storage facilities of substantial capacity to take the spent fuel from a number of power stations while waiting for reprocessing capacity to become available. Many different options for providing this away-from-reactor storage are being considered by the utilities operating nuclear power plants but final commitments to start building them had to await a clear government policy on waste management announced in 1983.

Germany

In the late 1950s and 1960s, German industry turned enthusiastically to nuclear power as an obvious source of energy to support a period of

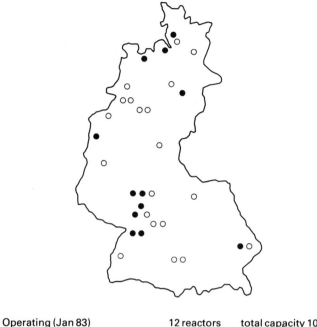

● Operating (Jan 83) 12 reactors total capacity 10 GWe
○ Under construction or planned 20 reactors total capacity 24 GWe

Figure 8.4 Nuclear power plants in West Germany

dramatic economic growth. Electricity is produced and distributed by a large number of private utilities, of which fourteen are mainly concerned with electricity generation and have traditionally purchased their plant and equipment from one of three private industrial groups. Nuclear research and development centres were established in a number of different federal states with complex funding arrangements from state governments and industry as well as the federal government.

In some respects the development of the nuclear industry in Germany has followed the pattern of development in the United States although the circumstances are clearly different. The industry, in association with the regional research centres and on its own initiative, took an early interest in the development of several different types of nuclear reactors and has pursued a number of these through to the prototype and demonstration stage. The final choices were dictated by the commercial judgement of private industry and the utilities rather than government. The Siemens company started out supplying pressurized water reactors under a Westinghouse licence with their main rival A.E.G. adopting the boiling water reactor under a General Electric licence. The German subsidiary of the Swiss Brown Boveri Company, which is the third traditional supplier of electrical generating equipment, attempted to enter the market with a British licensed advanced gas-cooled reactor and has followed this early interest in gas-cooled reactors with leadership of an industrial association which is continuing to develop the unique design of high temperature gas-cooled reactors with a pebble bed core. But the company won orders for two commercial nuclear power stations relatively late in the day by offering a pressurized water reactor in association with the German Babcock group and a licence from the American Babcock & Wilcox company.

After Siemens and A.E.G. had built a number of demonstration power plants of increasing size and obtained orders for the first full-sized commercial power stations with the rival reactor types, A.E.G. ran into severe financial problems which were solved by a merger of their heavy power engineering interest with Siemens into a joint company which was called Kraftwerk Union (K.W.U.). Siemens has since acquired full control of this company which now offers both types of reactor.

The German industry rapidly acquired independent capability in all branches of nuclear engineering and, shortly after its formation, K.W.U. was in a position to break away from the original Westinghouse licence and build the world's first 1300 MWe pressurized water reactor at Biblis. This plant has operated very successfully and with some evolution has led to a standard plant design which has been adopted for most of the subsequent nuclear power stations ordered in Germany. K.W.U. has also established a two-way technical exchange with Combustion Engineering who are building plants of a very similar design in the United States.

During the 1960s and 1970s the German industry demonstrated its ability to build nuclear power stations on schedule and to very high standards. With a very good operating performance, these plants have demonstrated a clear economic advantage over all other possible sources of electricity production in Germany, including domestic coal.

With a healthy domestic market and a number of export orders, K.W.U. seemed set to become a major force, along with the leading American companies, in world markets. The company was in fact relatively successful in obtaining orders in a number of different countries though some of these, notably in Iran and Austria, subsequently ran into difficulties for political reasons. In recent years the German nuclear industry has also suffered more than most from opposition to nuclear power in its domestic market. It has had to contend with endless legal actions brought by local groups and the associated delays of appeals to higher courts; it has had to face a complex political situation at the federal and state level which has relegated the use of nuclear power to a second option after domestic coal; and it has had to cope with licensing procedures which, in an apparent effort to respond to public concern, have become excessively demanding. One effect of this has been to add considerably to the cost of the standard design of German nuclear power plants and K.W.U. has lost some of its competitive edge, notably to the French, in the world marketplace.

There are signs of gradual improvement with political realization of the need for nuclear power to provide low cost electricity for the base load requirements of German industry. The situation at the start of the 1980s was that only eight of around twenty-five large nuclear power plants which had been planned during the previous ten years were actually operating, and the total capacity of nuclear generating plants was around 13 000 MWe instead of 20 000 MWe as had originally been envisaged. A further ten plants with a total capacity of 13 000 MWe were under construction and most of these should be coming on-line during the first half of the 1980s. A further twelve projects, with a combined capacity of around 15 000 MWe, were caught up in legal battles and some had been awaiting licensing approval since 1975. These were, however, starting to move into construction by 1982 and there are prospects that most of them will be operating by the end of the decade. This means that the nuclear contribution to electricity production in Germany could increase from a level of around 10 per cent in 1980 to 25 to 30 per cent by 1990.

German industry has established technical capability in all sectors of the nuclear fuel cycle, either independently or in collaboration with other countries. But as elsewhere introduction of all steps on a commercial scale has been hampered by politics.

Some small deposits of uranium have been discovered in Germany and a modest effort of exploration for further deposits is continuing. But as yet, only pilot scale production of uranium has been undertaken from domestic reserves. The main efforts of the German mining industry have been directed towards participation in joint ventures in other countries, such as Canada and Australia, with larger commercial prospects for uranium production. Some of the larger electric utilities have also taken a financial interest in such overseas mining activities in the interests of securing long-term supplies of uranium at predictable prices.

German industry is participating as a partner with the British and Dutch in the Urenco organization which is providing commercial enrichment services based on the centrifuge enrichment process. A centrifuge manufacturing plant in Germany has been supplying machines for one of the joint enrichment plants in the Netherlands and work has also started on the construction of an enrichment plant in Germany which should be put into operation by 1985. In addition, a novel enrichment process, known as the jet nozzle, has been developed in Germany and while it is not being adopted as the mainline effort to meet domestic requirements for enrichment, a small plant based on this technology is being built in Brazil as part of a package deal of nuclear power plants and supporting fuel cycle facilities.

From early work on several different types of reactor, the German industry has established the ability to fabricate most types of fuel. Pressurized and boiling water reactor fuel for the large commercial power reactors is now manufactured by a joint company – in which K.W.U. has the majority interest – formed from two organizations which originally held licences from the two American reactor manufacturers. An alternative source of supply for replacement fuel has also recently been set up in Germany by the nuclear subsidiary company of the American oil giant, Exxon (Esso).

While the nuclear industry was developing in Germany, it looked as if the traditional chemical companies would readily take up the challenge of providing commercial reprocessing capacity. When the question of reprocessing became a highly political issue with the prospect of long licensing lead times and poor profit margins, the chemical industry lost interest. The electric utilities, faced with a requirement to show that they had made adequate provision to deal with the spent fuel from their reactors, therefore formed a joint company called the German Reprocessing Company (D.W.K.) and has made provision for enough funding to ensure that the company can provide all the services needed at the, so-called, back end of the fuel cycle. This is similar to the course of action which has also been followed by electric utility companies in Japan, Sweden and Switzerland.

The first scheme put forward by D.W.K. was for a large integrated centre which would have included, at one site, facilities for interim storage of spent fuel, a large reprocessing plant and plants for conditioning all radioactive waste material for safe disposal in a repository constructed in a salt dome beneath the site. Politics at the state level prevented progress with this ambitious integrated project, which would have been large enough to deal with all the spent fuel from German nuclear power plants into the next century. Now alternative proposals for a number of interim storage facilities and smaller reprocessing plants in different states are being pursued. Decisions on licensing and which options are to be taken up are expected to be reached before 1985. In the meantime, geological investigations of the salt dome at the original site are proceeding to see if it is suitable for a waste repository which could be built in the 1990s.

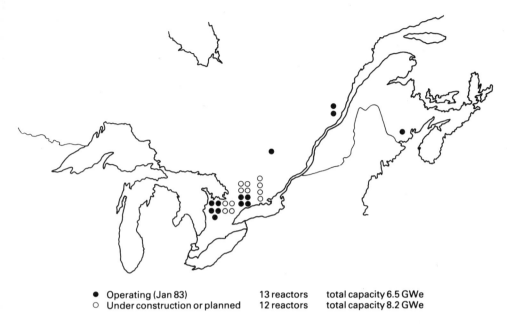

● Operating (Jan 83)   13 reactors   total capacity 6.5 GWe
○ Under construction or planned   12 reactors   total capacity 8.2 GWe

Figure 8.5 Nuclear power plants in Canada

Canada

Canadian experience with heavy water reactors dates back to 1942 when an important research reactor was built in support of the wartime effort to produce the first nuclear weapons. From this early beginning the Canadian industry has never deviated from the heavy water line in the development of civil nuclear power. Although some work has been done on variants with light water and organic cooling, the primary effort has been concentrated on the Candu system with heavy water moderator and coolant and natural uranium fuel.

A Crown corporation called Atomic Energy of Canada Limited (A.E.C.L.) has been mainly responsible for the development work, although in the early stages of commercial introduction there was an attempt by Canadian General Electric to provide a second design and construction organization to offer competition – this design team was eventually merged with that of A.E.C.L. Another important factor in the very successful development of the Candu system has been the close association of Ontario Hydro – the primary customer for nuclear power plant in Canada – with all stages of development. A utility with plenty of experience of large hydropower construction work, Ontario Hydro turned readily to nuclear power projects and was able to assume overall project management at construction sites.

Ontario Hydro has also had a lot to do with the very successful operation of the first two large commercial power stations at Pickering (with four 500 MWe reactors) and Bruce (with four 750 MWe reactors). These reactors have consistently dominated the top ten in worldwide tabulations of nuclear power plant operating statistics since they came on line during the 1970s – this despite the fact that the plants are relatively complicated. Since this success is partly attributed to the building of strings of four identical units, with the benefits this brings in plant familiarization and interchangeability of spares, the next stage of development is for four more 500 MWe units at Pickering and four more 750 MWe units at Bruce: these will be coming into operation during the mid-1980s. Work is now starting on four 800 MWe reactors to follow at a new site at the end of the decade.

Apart from this large programme of nuclear power plant construction in Ontario, the only other large commercial nuclear power plants in Canada are single 600 MWe reactors which have recently been completed in New Brunswick and Quebec. With this limited domestic market, A.E.C.L. has been fairly aggressive over the years in seeking export business. Some success has been achieved, notably in the Indian subcontinent and South-east Asia, and in South America. Additionally Romania has chosen to adopt the Candu system. But despite very active interest from other industrial countries in the good operating record and efficient use of uranium offered by the Candu system, it has so far failed to break into the markets of Europe and Japan which are dominated by light water reactors.

Assessment of the costs of electricity from nuclear and conventional sources is never easy because of the need to compare a system with a high initial capital cost but very low fuel costs with a system with lower capital cost and high fuel costs which are subject to unpredictable inflation. This difficulty is accentuated in the case of Candu because there is an additional capital cost associated with the inventory of the heavy water moderator while the natural uranium fuel costs are even lower than other nuclear fuel costs based on enriched uranium. In Canada, Ontario Hydro is able to make fairly accurate comparisons with large coal-fired power stations of comparable size and built at the same time as the main nuclear stations. Even making adjustments to the figures to allow for the fact that the nuclear stations are operated in preference to coal during periods of slack demand, the nuclear generating costs are significantly lower than coal. By the early 1980s the cost advantage was around 30 per cent and all the indications are that it will continue to increase.

If utilities in other countries could build and operate plants as efficiently as Ontario Hydro then there is a fair chance that a Candu reactor would show a marginal economic advantage over light water reactors. Decisions to adopt such a system are, however complicated by the higher initial cost, by consideration of the amount of worldwide experience of the system, and by comparison of the alternatives of establishing either heavy water or enriched uranium supplies.

Heavy water production was in fact the area where the Canadian industry encountered most problems in the early days. Although the basic technology of a suitable chemical process was established without much difficulty, scaling it up to a commercial plant of comparable size and complexity to a petrochemical plant was not easy. One early industrial project ran into so many difficulties that it practically had to be rebuilt from scratch. Finally Ontario Hydro got on top of the problem by building an impressive series of heavy water production units at a site adjacent to the Bruce power station, but there was a difficult period when the start-up of some of the new reactors was in jeopardy.

Now the heavy water situation has been turned around. In contrast to uranium enrichment services where there is a steady demand throughout the life of a power station, a heavy water reactor has a large start-up requirement to provide the inventory in the moderator tank but then only has a small demand to make up for minor losses. Canadian industry has therefore been faced with the difficult task of having to try to match heavy water production capacity to the anticipated rate of building new power stations with Candu reactors. With nuclear power plant ordering worldwide lower than had been expected, there is likely to be a substantial overcapacity for heavy water production throughout the 1980s and two Canadian plants are presently mothballed.

Canada is the second largest producer of uranium in the non-Communist world, after the United States. The uranium mining industry was established with early exploration and mining to meet military demand in the 1940s but this was followed by a period of recession, accentuated by the American embargo on uranium imports, until the worldwide market for uranium to fuel civil nuclear power programmes was established in the 1960s. There was a difficult political period in the early 1970s when export policy and safeguards against possible misuse of Canadian uranium came under scrutiny. But this has been largely overcome with lengthy negotiation of new agreements with Europe and Japan – the main customers for Canadian uranium. The government has adopted a policy of earmarking adequate uranium resources to meet the lifetime fuel requirements of foreseeable nuclear power plants in Canada but there is plenty left over to support a major export industry. In common with other mineral activities in the country, foreign participation is virtually unrestricted at the exploration stage but once mining production starts there is a requirement for no more than one third foreign participation in the operating company.

The domestic requirement for nuclear fuel is, of course, confined to natural uranium. From time to time there has, however, been discussion of the possibility of building uranium enrichment plants in Canada which would effectively use cheap hydro and nuclear power to give added value to the uranium before it is exported. But with a likely worldwide surplus of enrichment capacity well into the 1990s, any such projects have been pushed far into the future.

Japan

There is probably no country in the world that has a stronger case for the adoption of nuclear power than Japan. There are practically no indigenous sources of energy and certainly no prospects of supporting a very large population with a rapidly rising standard of living and a dramatic rate of economic growth, without heavy dependence on imported fuel. During the 1970s the dependence on imported oil and coal reached the staggering level of 90 per cent and even with a large effort to conserve energy and a large nuclear power plant construction programme, this dependence is only likely to be reduced to 60 per cent by 1990. The country also lacks most other raw materials and is heavily dependent on maintaining its very competitive position for manufactured goods in world markets if it is to pay for raw material needs. Nothing poses a greater threat to this competitive position than the rising cost of oil imports.

The suitability of nuclear power to meet the country's future energy needs, particularly for electricity, was recognized at an early stage. In the late 1950s Japan ordered a gas-cooled reactor of the Magnox type from Britain, shortly after it had started work on the first programme of commercial nuclear power. This was closely followed by the ordering of a small demonstration plant with a boiling water reactor from the United States. These were the first two nuclear power plants that had to be constructed to withstand the high probability of seismic shocks which exists in Japan.

At the same time a large nuclear research and development effort was mounted in Japan and has been maintained over the years, with a close collaboration between government agencies and private industry. This has ensured the development of a complete range of competence in nuclear technology in Japan. But an interesting approach, closely based

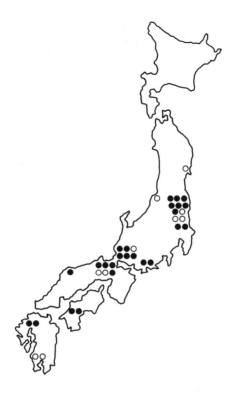

| ● Operating (Jan 83) | 25 reactors | total capacity 17 GWe |
| ○ Under construction or planned | 10 reactors | total capacity 8 GWe |

Figure 8.6 Nuclear power plants in Japan

on American technology, was adopted by the seven private electric utilities when they started to build large commercial nuclear power stations during the 1960s. For the first unit of any particular reactor type, and also for the first unit of a particular size, the prime contact was placed with American companies, Westinghouse or General Electric, and construction was carried out with one of the three major Japanese engineering groups – Mitsubishi, Hitachi and Toshiba – acting as major subcontractors. For subsequent orders of similar designs of reactors the Japanese companies have assumed the role of prime contractors with licences from the American companies.

This has ensured rapid transfer of new technology but it has also meant that any problems encountered in the United States, both of a technical kind and those associated with public opposition to nuclear power, were also transferred quickly to Japan. The authorities in Japan have tended to react strongly to any such problems, whether real or perceived, and have imposed very stringent controls on the industry. During the late 1970s this resulted in quite a lot of delay in the licensing of plants, demands for a large amount of backfitting to existing plants and a rather arbitrary requirement for nuclear plants to be shut down for extended inspection each year, all of which has contributed to Japanese nuclear plants appearing to have been less productive than comparable plants in other parts of the world.

Nonetheless, the government and the utilities remain firmly committed to a large programme of nuclear power plant construction as the only practical means of substituting for increasingly expensive imports of coal and oil. During the 1980s the nuclear generating capacity is expected to increase from a level of 15 000 MWe, producing just over 10 per cent of the country's electricity, to around 40 000 MWe, meeting about 25 per cent of electricity demand. This puts Japan a close second to France in the size of construction programme during the decade.

Somewhat surprisingly the Japanese nuclear industry has not so far won any export orders for nuclear power plants. The three industrial groups have clearly established adequate independent capability, one with pressurized water reactors and two with boiling water reactors, to build plants in other countries and they have been joining in international bidding for a number of prospects. If the market for nuclear power plant around the world picks up in the coming years it will probably only be a matter of time before the Japanese break into the market along with the suppliers from France, the United States, Germany and Canada. In the meantime, Japan's heavy engineering companies have certainly established themselves in the world market for the supply of components for the nuclear industry, notably with large steel forgings for pressure vessels and with complete pressure vessels.

As well as being devoid of fossil fuel resources Japan has no uranium. This is less of a problem because nuclear power costs are relatively insensitive to uranium costs. Nonetheless, the Japanese industry has shown concern to ensure long-term supplies of uranium for the country's large nuclear power programme. To this end the industry has entered into a number of joint mining ventures in several major uranium producing areas, notably Australia, Canada and central Africa. The prospect of uranium supplies becoming a political issue is also viewed with concern, especially following the Carter era in the United States during which Japan encountered some difficulties and delays in obtaining consents for transport and reprocessing of spent fuel of American origin. As a backstop solution to possible future problems some development work has been undertaken on methods of extracting uranium from sea water. This is a potentially unlimited source of supply but the concentrations are so low that the cost of recovery would be prohibitive for use in present-day thermal reactors except under circumstances of extreme limitations on the free world market.

Similar concern about future political developments has provided an incentive for Japan to strive for independent capability in all sectors of the nuclear fuel cycle. A state-owned organization known as the Power Reactor & Nuclear Fuel Development Corporation (P.N.C.) is mainly responsible for development work in this area. Although Japanese utilities have acquired long-term contracts for uranium enrichment services from the United States and Europe to cover the needs of the nuclear power plants built through to the 1990s, P.N.C. has devoted a large effort to independent development of the centrifuge enrichment process. A pilot plant with a capacity of 50 000 separative work units (S.W.U.) per year is already operating, a demonstration scale plant with a capacity of 250 000 S.W.U. is due to be operating by the mid-1980s, and a full scale commercial plant of 1 million S.W.U. is planned for the late 1980s.

Likewise, although Japanese utilities have reprocessing contracts with Britain and France that will cover all spent fuel arising during the 1980s, there is also an interest in developing an independent capability in this field. P.N.C. is now obtaining operating experience with a pilot scale reprocessing plant of about 200 tonnes per year capacity which was built with French technical assistance. The next planned stage is a commercial scale plant of about 1000 tonnes per year which a joint utility company is hoping to build, with the participation of the chemical industry, for operation in the 1990s. The biggest problem, as in Germany, is going to be finding a site for such a plant which is not resisted by nuclear opponents.

The rest of Western Europe

Special mention should be made of Sweden which, quite independently of any licensing arrangements from the leading nuclear nations, has developed a very successful boiling water reactor. The company responsible for this achievement is A.S.E.A. which was already well-established as a supplier of electrical engineering equipment in international markets. A state research and development organization, A.B. Atomenergi, after early efforts to develop a unique line of heavy water reactor, subsequently joined forces in an organization called A.S.E.A.-Atom which has been responsible for seven completed plants and two further plants due for completion in the mid-1980s. A further three

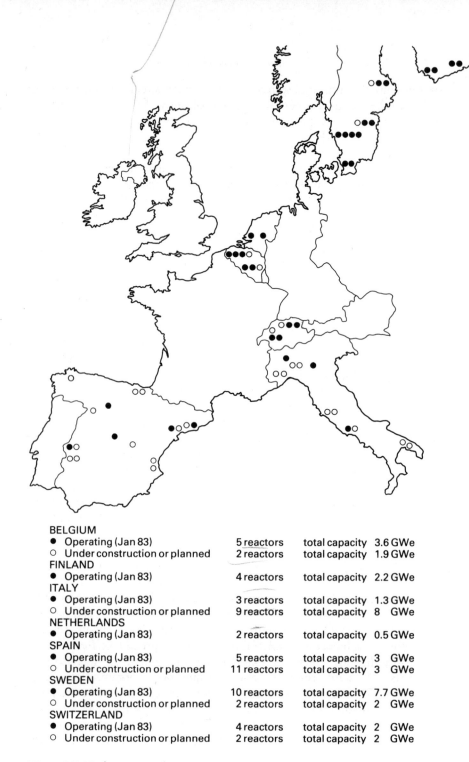

Figure 8.7 Nuclear power plants in Western Europe

plants equipped with pressurized water reactors supplied by a European subsidiary of Westinghouse complete a national programme of twelve nuclear reactors which, following a referendum in 1979, has been set as a rather arbitrary limit on the use of nuclear power in Sweden. A.S.E.A.-Atom has also successfully completed two plants in Finland and might well have had further successes in the export market but for the political turmoil at home surrounding a heated five-year debate on the issues of nuclear energy.

Despite the political difficulties, nuclear power is making a very important contribution to electricity production in Sweden. In a country which has a *per capita* consumption of electricity which is among the highest in the world, the contribution of nuclear generation is over 30 per cent and this should increase to more than 50 per cent by the mid-1980s. The problem for the industry is to maintain its independent capability to build nuclear power plants after completion of the last two reactors in the twelve-reactor domestic programme. A.S.E.A.-Atom is still bidding actively in the international markets but, even if successful, the volume of business is unlikely to support the company for long unless there is a reconsideration of the domestic policy. The best hope is that towards the end of the decade, continued reliable operation of the Swedish nuclear power plants, and the low cost electricity that they produce, will demonstrate to politicians and the public that the nuclear option is both desirable and acceptable.

Italy might well have become a major nuclear industrial country if it had not been for years of political indecision on the implementation of a programme of nuclear power plant construction. At one time it was envisaged that new projects should be started at the rate of around one new power station each year throughout the 1970s. In the event, the country will be lucky to achieve half this target during the 1980s. The electricity supply system is now under severe strain to meet winter demands, and dependence on oil imports is the highest in Europe.

This rather sad story follows very early recognition of the potential role of nuclear power in meeting energy needs in a country with no fossil fuel resources, and most of the sites for hydropower already exploited to the full. In the late 1950s, when electricity production in Italy was undertaken by private utilities, three early nuclear power plants were ordered – one gas-cooled reactor of the Magnox type from Britain and pressurized and boiling water reactors from the United States – and Italian industrial groups established technical ties and licensing arrangements to allow them to undertake design and manufacture of future reactors and the supporting fuel fabrication services. By the time the three early plants were completed the utility industry had been nationalized. Although the state utility, known as E.N.E.L., continued to show great enthusiasm for nuclear power it has only been able to complete one further boiling water reactor and make a start on the construction of two more units. The industrial groups with interest in nuclear design and construction have undergone numerous reorganizations and have finished up as one group of mostly nationalized companies. For the latest national energy plan the decision has been taken to concentrate effort on one design of reactor and, after some fairly intense argument, the industry has settled upon the pressurized water reactor. Progress has been made recently with the difficult problem of finding sites for some four new 1000 MWe plants which it is hoped will be built and put into operation by 1990. If this is achieved the contribution of nuclear energy to Italy's electricity production will be increased from 3 per cent to around 15 per cent during the decade.

Despite the problems with the country's nuclear power programme, Italian scientists and engineers have taken readily to the challenge of nuclear technology. A national research and development organization has undertaken impressive work on a wide range of different projects and the manufacturing industry has been moderately successful in selling high quality components for nuclear plants in many parts of the world. Not least of the industry's activities are its participations with the French in the building of the Super-Phenix fast reactor and in the Eurodif gaseous diffusion enrichment plant.

Belgium is anther country that recognized at an early stage that nuclear power could play a very important role in a highly industrialized country with very few indigenous sources of energy. In spite of a good deal of political uncertainty in government in recent years, the nuclear power programme has been allowed to proceed relatively smoothly. By 1980, four nuclear reactors were supplying around 20 per cent of the country's electricity and another four, at an advanced stage of construction, were scheduled to increase this to over 50 per cent by the mid-1980s. But since Belgium is dependent on expensive imports for the rest of its energy requirements, there is now pressure from the private utility industry to obtain government decisions on further nuclear power plants for the future.

As a result of a research and development programme which was fairly substantial for such a small country, the Belgian nuclear industry has established a strong technical potential in a number of specialist fields, notably the nuclear fuel cycle and more particularly in the fabrication of fuel assemblies containing plutonium. The heavy engineering industry has also turned its hand to the fabrication of major components for nuclear projects in a number of countries.

But due to the size of the country, international collaboration is the name of the game for Belgium. The commercial nuclear power plants have been built in collaboration either with the French company, Framatome, or with the European nuclear subsidiary of Westinghouse, which is based in Brussels. For advanced reactor development Belgium – together with the Netherlands – is a partner in a prototype fast reactor project in Germany. An interest in uranium enrichment has been provided with participation in the Eurodif gaseous diffusion plant in France. From 1966 to 1976 a small reprocessing plant was operated in Belgium by an international company called Eurochemic which had been established by a number of countries under the auspices of the O.E.C.D. European Nuclear Energy Agency. Ownership of this facility has now been transferred to Belgium and the government has decided that it should be reactivated. When this is done it may well be accompanied by a reciprocal arrangement with France, which has contracts for the reprocessing of the bulk of spent fuel from the standard commercial power plants in Belgium, but could be interested in access to the smaller flexible plant in Belgium to deal with the spent fuel from research and test reactors.

In Spain, privately owned utilities took an early initiative in purchasing two plants equipped with pressurized and boiling water reactors from the United States and later bought a gas-cooled reactor from France, all of which have operated very successfully. But the main commercial nuclear power programme was launched in the early 1970s when the utilities got together and ordered a batch of six standardized pressurized water reactors, each of 900 MWe capacity, from the European subsidiary of Westinghouse. An important feature of this deal was an increasing content of Spanish subcontracts and on the strength of the programme, the Spanish nuclear industry has established impressive manufacturing facilities for major components.

The first of the 900 MWe plants are now coming into service after several years of delay in construction, but one station in the Basque region has been the target of terrorist attacks and is suffering further delays and uncertainty about its future. Since the batch of six standard plants, a further nine units have been ordered – some more from Westinghouse as well as boiling water reactors from the American General Electric company and pressurized water reactors from Germany's Kraftwerk Union. The original idea was that these plants, coming into operation during the 1980s, would make an important contribution to reducing oil imports and that the contribution of nuclear generation to the country's electricity supplies will increase from 5 per cent to around 40 per cent during the decade. But the Socialist government which came to power in 1982 has expressed some reservations about the size of the nuclear power programme, especially in the light of a slow-down in the growth of demand for electricity, and there is a possibility that some of the projected plants will be halted or delayed.

In Switzerland, traditional experience from large hydro-power projects has led to the establishment of an electrical engineering industry with a worldwide reputation. Availability of hydropower has also been an important factor in establishing a high standard of living in a country devoid of any other energy resources. Commercial nuclear power came along at about the time that all the easily exploitable sites for hydro projects had been developed, and the private utility industry turned enthusiastically to the new energy sources as a means of meeting growth in demand for electricity without having to pollute the famous Swiss air by burning fossil fuels. Two nuclear power stations – one with two pressurized water reactor units and the other with a boiling water reactor – were built in the late 1960s by partnerships between American reactor suppliers and the Swiss Brown Boveri Company (B.B.C.). These two projects provided a striking example of what can be achieved with good management. Each unit was constructed in less than four years and has operated very well.

But Switzerland was the first country in Europe to import public opposition to nuclear power from across the Atlantic, and subsequent projects have encountered severe delays in getting authorization for construction. One plant has been effectively blocked for more than twelve years and is still the centre of a major storm which could decide the ultimate fate of the nuclear power construction programme in Switzerland. Of two other projects, one pressurized water reactor supplied by Germany's Kraftwerk Union was successfully completed and put into operation in 1980 and the other, a boiling water reactor being built by the same Swiss-American partnership as before, is due to operate by 1984–85. This plant should increase the contribution of nuclear generation of electricity in Switzerland from the present level of 20 per cent to nearly 30 per cent, but beyond that the future is very uncertain.

The Swiss government has won narrow support for a continued programme of nuclear power plant construction as long as the projected need for the electricity is established and satisfactory provisions are made for waste management. But these conditions provide a source for endless argument and long delays. Swiss industry has, in the meantime, been making a significant contribution in world nuclear markets, both with the supply of high quality components for a variety of different types of reactors and also with engineering consultancy services for countries planning to embark upon nuclear power programmes.

Finland has provided a testing ground for comparison of Western and

Eastern nuclear technology. A power station with two pressurized water reactors supplied by the Soviet Union is being operated by a state-owned electricity utility, while a group of private utilities is operating two boiling water reactor units supplied by Sweden's A.S.E.A.-Atom at a second station. So far, however, comparisons are a little inconclusive – both projects have had their fair share of delays during construction but since starting operation 1977 and 1979 the first units, at least, have been performing well. Finnish industry, with quite a lot of applicable experience in quality engineering from the country's paper and shipbuilding industries, made a significant contribution in fabrication of equipment for both power stations, including a containment building for the Soviet reactors which are of a type that does not normally have one in Eastern block countries.

Finland's four operating reactors supply a surprisingly large 25 to 30 per cent of the country's electricitiy requirements and have helped in the achievement of a reduction in oil imports during a period of growing total demand for primary energy. However, no new projects have yet been put in hand, so further improvements in this direction will not be possible for some years. The state-owned utility has been talking about plans to build one or two more pressurized water reactors, each with a capacity of 1000 MWe, for several years and has been looking at designs offered both by the Soviet Union and France's Framatome. There is also considerable interest in the possibility of using waste heat from future nuclear power plants to feed into the large district heating network of the Helsinki metropolitan area. But while there is no great dissent on the part of the public, neither is there great urgency on the part of government to take decisions on these future plans.

For the rest of Western Europe, the Netherlands and Austria have been most active in the nuclear field with involvement in research and development projects both nationally and internationally from the early days. The manufacturing industries in both countries have also been able to apply traditional skills in high quality engineering to the supply of components for nuclear projects in many parts of the world. A very small demonstration boiling water reactor with a capacity of 50 MWe has been operating in the Netherlands since 1969 and a larger 450 MWe pressurized water reactor, supplied by the German company Kraftwerk Union, was put into operation in 1973. But at the time that both countries were ready to launch further programmes of nuclear power plant construction to cut their heavy dependence on oil imports, they ran into strong public and political opposition to nuclear energy. This has prevented the Austrians from putting into operation a completed 700 MWe boiling water reactor supplied by Germany, while in the Netherlands there has

Nuclear Energy

been considerable pressure for closure of the operating plants and protracted delay of a debate to decide about a future nuclear power plant construction programme.

Although both countries are relatively small they have well-established grid distribution systems for electricity and the ability to make use of interconnections with neighbouring countries which allows them to balance short-term surpluses and shortages of supply with exports and imports of electricity. This, together with their strong industrial base, means that they could accommodate large economic nuclear power stations and just one or two such stations would be able to make a major percentage contribution to national electricity generation with corresponding benefits in lower cost electricity for industry and reduction of the balance of payments burden of oil imports. There are some signs of slow realization of these factors by political leaders and, following further public debates in the coming years, there is a prospect of revival of nuclear power programmes in the second half of the 1980s.

In several other West European countries, such as Denmark, Greece, Ireland, Norway, Portugal and Turkey, plans to introduce nuclear power have been studied and discussed many times but for a variety of reasons, usually political but in some cases also financial, decisions have been postponed. There are signs that decisions to proceed with first nuclear power plant projects in some of these countries may be taken in the next few years and, depending on recovery from the effects of worldwide recession, there may be a considerable revival of interest by the end of the decade.

The Developing World

The larger developing countries of the world have shown a keen interest in nuclear power throughout the years of civil development and several have already established a sound base of research and development and the beginnings of their own nuclear industries. In principle nuclear power could make a major contribution in meeting a rapid expansion of energy demand which will be essential to support the programmes of industrialization that they are all seeking to implement and also to support the needs of rapidly expanding populations. In practice the introduction of nuclear power has proceeded rather more slowly than might have been expected. This is due to a variety of reasons which include problems associated with the ability of the electricity distribution systems to accommodate large commercial nuclear power stations, the need to establish an infrastructure for administration and control of nuclear energy developments and, above all, the problems of raising

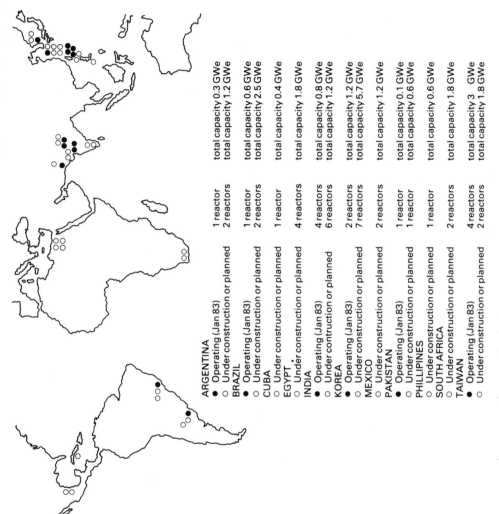

Figure 8.8 *Nuclear power plants in the developing world*

finance for capital-intensive nuclear projects in competition with the demands of many other development programmes.

The developing world has seen little if any of the public opposition to nuclear power that has been experienced in advanced industrial countries. The concerned citizens of the affluent middle classes who have dominated the anti-nuclear movements in the industrialized world do not exist in developing countries, but concern for improvements in living standards certainly does. On the other hand the introduction of nuclear power has suffered just as much from politics and bureaucracy in the decision-making process. The inevitable need for expansion of energy supplies and the crippling costs of oil are, in the end, likely to prevail over the early problems and in the coming years we could see a dramatic growth in the use of nuclear power, especially to meet the demands of the huge new urban conurbations which, like it or not, are growing up in South America and South-east Asia.

India took an early initiative with the setting-up of an impressive research and development centre and followed this with the building of two power stations. The first, equipped with twin boiling water reactors of 200 MWe capacity supplied by the United States, has been in operation since 1969. Two heavy water reactors, each of 200 MWe capacity, were supplied by Canada for the second station and started operation in 1973 and 1981. These were intended as the first of a programme of similar units in which the Indian Department of Atomic Energy would take over from the Canadians as the prime contractor, and Indian industry would play an increasing role in the manufacture of component plant and equipment. At the same time the Indians were establishing complete capability to provide the fuel cycle services for the natural uranium-fuelled reactors, including fuel fabrication from domestic uranium resources, heavy water production and reprocessing.

The Indian test explosion in 1976 of a nuclear device, ostensibly for peaceful applications such as excavation of irrigation schemes, caused a worldwide furore and resulted in Canada terminating all technical assistance on the power station building programme. The Indian industry has continued construction on its own, albeit at a slower pace, and the next four units are due to start operating during the mid-1980s.

Compared with the huge population that it has to support, the large land mass of India contains remarkably few energy resources other than uranium – and thorium which can be converted efficiently into nuclear fuel in the heavy water type of reactor. There have therefore been frequent political declarations of intent to undertake a large programme of nuclear power plant construction, probably involving ten more heavy water reactor units, in the coming years. Now that the independent capability of India's nuclear industry to build and operate plants is near to realization, it is likely that this programme will get under way.

The most ambitious programme of nuclear power plant construction in the developing world has been undertaken by South Korea. This started with a 560 MWe pressurized water reactor supplied by the United States which was put into service in 1978, and a second unit of 600 MWe which is due to operate in 1983. A 600 MWe heavy water reactor from Canada is also nearing completion and a further six pressurized water reactors of 950 MWe from American and French suppliers are under construction. When all these plants come on line in the second half of the 1980s and the early 1990s they will account for over 40 per cent of the country's electricity production. Rapid growth in electricity demand is accompanying a dramatic programme of industrialization in South Korea and with few domestic energy resources the country is looking mainly to nuclear energy to provide a large measure of independence from imported oil. There is talk of increasing the total nuclear generating capacity to around 80 000 MWe by the year 2000 when it will be supplying 20 per cent of the primary energy requirements. While relying heavily on imported technology to get this nuclear power programme launched, the Koreans are striving to achieve a progressively larger domestic work content and clearly look to the day when they will be able to undertake the bulk of the manufacture and construction themselves.

Taiwan is another country which, with no domestic energy resources, is putting a heavy emphasis on nuclear power as a means of supporting a programme of industrialization. Already three American-supplied boiling water reactors, two of 600 MWe and one of 950 MWe, are operating and a fourth boiling water reactor and two pressurized water reactors are under construction and due to be operational by 1985. The three operating plants already account for 12 per cent of electricity generation and the three follow-on units will increase this to around 40 per cent. This means that both in total capacity of nuclear plant and percentage of electricity production, Taiwan is currently leading the developing world in use of nuclear energy.

Latin American countries with their large and expanding populations and correspondingly rapid growth in the demand for energy, represent another situation where there is a strong case for nuclear energy. So far three countries, Argentina, Brazil and Mexico, have embarked on nuclear power programmes but the rate of decision-making and implementation is relatively slow and has involved quite a lot of politics both at national and international levels.

Argentina adopted a novel design of heavy water reactor from Germany for its first nuclear power plant. It makes use of a large pressure vessel similar to those of light water reactors, rather than the pressure tube arrangement used in the Canadian design of heavy water reactor. This first plant, with a capacity of 300 MWe, has operated exceptionally well since it went into service in 1974. A second unit of this type, but with the capacity increased to 700 MWe, has been ordered from Germany for construction at the same site. In the meantime, however, a 600 MWe Candu-type of heavy water reactor has been ordered from the Canadians and is due to be completed in 1983. Despite rather protracted and political negotiations with the Canadians over the contractual arrangements for the plant that they are supplying, the Argentinians have indicated a technical preference for the pressure tube design of reactor in the future.

Brazil has opted for the pressurized water reactor for its nuclear power programme. A first 600 MWe unit was supplied from the United States and, after some construction delays, was finally put into service in 1982. But the main Brazilian programme is based on a large package deal concluded in 1975 with Germany. This includes two 1200 MWe units which, also after some delays, are due to be completed in the second half of the 1980s and options to purchase a further four identical units. The schedule for these follow-on units has become a little uncertain, partly because of some public opposition to nuclear power but also because the dramatic rate of growth in electricity consumption which was being experienced in the mid-1970s has slowed down.

A feature of both the Argentinian and the Brazilian nuclear power programmes is an effort to acquire industrial capability to provide the supporting fuel cycle services domestically. This has been interpreted in some quarters, notably by the media and nuclear opponents, as a devious effort to acquire nuclear weapons capability. In fact the decisions date from a period when the advanced industrial countries were expressing concern about the timely availability of commercial capacity for uranium enrichment and reprocessing to meet the demands of their projected programmes of nuclear power plant construction. Even when this fear of shortage was turned around, the politics of the Carter era placed severe doubts upon the reliability of nuclear fuel cycle services from both the United States and Canada. A less suspicious interpretation of the construction of fuel cycle plants in Argentina and Brazil is, therefore, a desire to make sure that they have the independent technical skills to fuel their vital power-producing reactors throughout their working life. All the nuclear power plants and fuel cycle facilities are covered by safeguards agreements with the International Atomic Energy Agency and in most cases by more stringent bilateral agreements with the supplier countries.

Mexico has two 650 MWe boiling water reactors, supplied by the United States, under construction and due to be in operation by 1985 after some years of delay. This was due to be followed by a steady programme of nuclear power plant construction which has stimulated a lively round of international bidding from the major supplier countries. But a severe financial crisis in 1982 forced the government to postpone immediate placing of orders for new plants. The declared intention is still to proceed with a large programme when the financial situation is better.

In the rest of the developing world there have been a number of small beginnings in the use of nuclear power, such as a 125 MWe heavy water reactor in Pakistan and a 600 MWe pressurized water reactor in the Philippines, and many studies of prospects have been carried out in other countries. Prior to the 1973 oil crisis most studies indicated that the case for nuclear power in the smaller developing countries was marginal. It would depend partly on the development of a standardized small size power reactor more suited to the needs of countries without well-established grid distribution systems, and also on substantial assistance from advanced countries in training and finance. The rapid escalation in world oil prices has hit the non-oil-producing countries of the developing world particularly hard and now the economic case for nuclear power is much stronger. But the raising of finance has also become more difficult; still more importance is being placed on the need for operator training; and, while there are a number of promising small reactor designs on paper, none is yet established as a standardized product. Many informed observers believe that there will eventually be a breakthrough in the use of nuclear power in the developing world but few are willing to predict precisely when and how this will happen.

The most immediate future prospect is in Egypt where a large nuclear power programme is now being planned. This could result in the construction of eight 1000 MWe plants during the late 1980s and 1990s. Agreements have been concluded with potential supplier countries and commercial contracts are likely to follow in the coming years. Israel also has an active interest in building nuclear power stations for both electricity production and water desalination but the question of supply is likely to be dominated by the political question of acceptance of safeguards controls on all nuclear installations in the country. Finally, in Iran there is some revival of interest in completing construction of at least some of the nuclear power plants which were started under the ambitious programme of the former Shah, but the extent to which the supplier countries, Germany and France, are prepared to become involved again is rather questionable.

Comecon countries

Although the Soviet Union lays claim to having operated the world's first nuclear power station – a very small 5 MWe plant started up in 1954 – the large-scale commercial exploitation of nuclear power did not receive a high priority in the development plans of the 1950s and 1960s. Rather the emphasis was on the development of a broad base of nuclear technology with the main interest for commercial power production centred on the eventual development of the fast reactor with its efficient use of what were, at the time, thought to be limited uranium resources. The design of reactor which had originally been developed for military production of plutonium was, however, used to demonstrate the parallel production of electricity. This reactor has a graphite moderator and boiling light water cooling of low enrichment fuel in pressure tubes. As such it is modular in nature and has been built in progressively larger sizes with electricity production taking over as the primary role.

As uranium availability and the commercial prospects of thermal reactors improved in the early 1960s, a second line of pressurized water reactor development was also followed and by the end of the 1960s a standard 440 MWe unit was being installed in a number of power stations both in the Soviet Union and in other Comecon countries.

In the development plans for the 1970s nuclear power was starting to receive a greater emphasis but still not enough to overcome the bureaucracy which tends to dominate production priorities in Soviet manufacturing. As a result the total capacity of operating plants in 1980 was only 10 000 MWe and fell a long way short of the 30 000 MWe target which had been set at the start of the decade.

Most of the manufacturing work for both the graphite moderated pressure tube reactor and for the pressurized water reactor components was carried out at a large traditional engineering works in Leningrad. This has been able to cope well with the modular components of the pressure tube reactors which have been built to schedule in unit sizes up to 1000 MWe and have generally operated very successfully. But the factory was hard pressed to meet schedules for the larger components of the pressurized water reactors, especially in a 1000 MWe size which was originally scheduled to start coming into service in 1977 but did not finally make it until 1982. Building of an impressive special-purpose factory with a capacity to manufacture up to eight sets of components for 1000 MWe pressurized water reactors each year, was undertaken in the mid-1970s with technical assistance from the Breda company in Italy. Known as Atommash, this factory located at Volgodonsk is coming into operation about two years behind schedule and should greatly improve the situation when it reaches its full capacity around 1985.

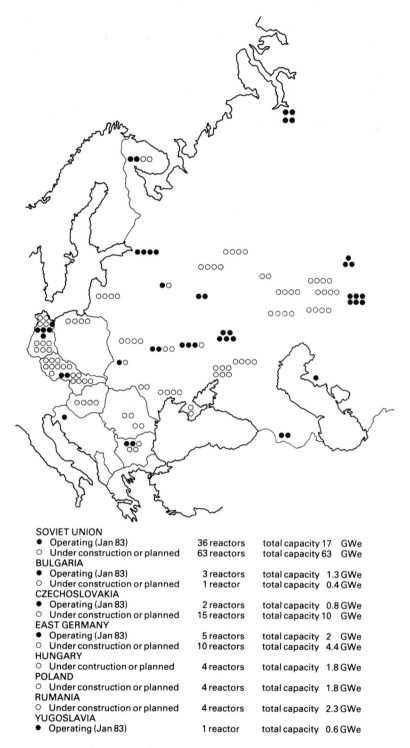

Figure 8.9 Nuclear power plants in Eastern Europe and the Soviet Union

Nuclear Energy

This will be none too soon because by the end of the 1970s the need for nuclear power had assumed a top priority in the planning of the Comecon countries. At a summit meeting in Warsaw in 1979 a bold ten-year programme calling for an increase in the total installed nuclear capacity to 150 000 MWe by 1990, was agreed. Of this total some 115 000 MWe is planned for the European region of the Soviet Union to the west of the Urals – accounting for 60 per cent of the population – and 35 000 MWe is envisaged in other Comecon countries. In an effort to get on top of the manufacturing of components for this ambitious programme, agreement was also reached in Warsaw to share the work on standardized pressurized water reactors among east European countries. In particular this programme will make use of the Skoda works in Czechoslovakia which has an established reputation for the manufacture of large, high quality, steel components.

Most observers are, with some justification, a little sceptical of the Comecon countries achieving their 150 000 MWe target by 1990. But with the political priority now being given to nuclear power and with the help of a widely-based manufacturing effort on highly standardized plants, they may come closer to meeting their projections than the nuclear industry in the West.

Glossary

A.G.R. – Advanced Gas-cooled Reactor – a gas cooled reactor using enriched uranium fuel.

ALPHA PARTICLE – emitted during the radioactive decay of some atoms and originally referred to as alpha rays. It is made up of two neutrons and two protons. It has a positive electrical charge and a relatively large mass compared with other nuclear particles.

ATOM – the basic building block of all matters. An atom has a relatively heavy nucleus made up of positively charged protons and neutral neutrons surrounded by orbiting electrons with a negative charge balancing that of the protons in the nucleus. The number of electrons (from 1 to 92) determines the chemical characteristics of the atom. The number of neutrons and protons (from 1 to 238) determines the weight and isotope of the atom.

ASSEMBLY – (of nuclear fuel) takes the form of rods or elements, usually mounted into arrays of up to 300 to go into a reactor. A typical reactor core could contain several hundred fuel assemblies.

BACKGROUND RADIATION – the naturally occurring nuclear radiation coming from outer space as cosmic radiation, or from naturally occurring radioactive atoms in the materials of the earth.

BECQUEREL – (Bq) the unit now used for specifying amounts of radioactivity, equal to the number of atoms disintegrating per second. The former unit was the Curie = 3.7×10^{10} Bq.

BENTONITE – Clay with strong absorbing properties likely to be used for back filling waste repositories.

Glossary

BETA PARTICLES – also sometimes referred to as beta rays. These are negatively charged electrons emitted during the radioactive decay of atoms. They are much lighter than alpha particles.

BINDING ENERGY – the energy associated with the forces that hold nuclear particles together in the nucleus of an atom. Some of the binding energy is released when heavy atoms are fissioned or light atoms are fused together.

BLANKET – non-fissile uranium-238, placed around the core of a fast reactor in a form similar to fuel assemblies, in order to absorb the surplus fast neutrons escaping from the core. The neutrons convert it into a useful fuel, plutonium.

BLOWERS – fans used to circulate the gas coolant around a reactor.

BN-600 – the large fast reactor with an output of 600 MWe operating at Beloyarsk in the Soviet Union.

B.N.F.L. – British Nuclear Fuel Limited – state-owned fuel cycle company.

BREEDING – the process of converting uranium-238 into plutonium in a blanket around a reactor core, usually a fast reactor, to produce additional plutonium fuel.

B.W.R. – Boiling Water Reactor – a light water reactor with enriched uranium fuel and boiling water coolant.

BURN-UP – the degree to which uranium, and plutonium, atoms in nuclear fuel have been consumed by the fission process in a reactor.

CANDU – CANadian DeUterium reactor – the heavy water moderated and cooled reactor with coolant pressure tubes, developed in Canada.

C.D.F.R. – Commercial Demonstration Fast Reactor – a projected first-of-a-kind fast reactor with a power of 1200 MWe proposed for Britain.

CANNING – (also called cladding) – the metal sheath in which nuclear fuel is sealed to form fuel elements. It allows heat generated during fission to be conducted to the coolant but seals in the radioactive fission products.

C.E.A. – (Commissariat à l'Energie Atomique) – the state research and development organization in France.

Glossary

C.E.G.B. – (Central Electricity Generating Board) the state-owned utility responsible for electricity production in England and Wales.

CENTRIFUGE – a very high speed spinning drum used to separate heavy atoms from lighter atoms by centrifugal force in the uranium enrichment process. It works like a spin dryer.

CHAIN REACTION – the process in which neutrons released during the splitting of one uranium atom go on to split other atoms, creating a self-sustaining reaction.

CHARGED PARTICLE – parts of an atom possessing a small charge of static electricity. A proton has a single unit of positive charge. An electron has a single unit of negative charge. A very large number of electrons passing through a conductor creates an electrical current.

CHERENKOV RADIATION – named after the Russian scientist who first observed it. It is the light emitted when a charged particle moves at a speed greater than the speed of light in a material such as water or glass. The speed of light in such material is lower than it is in a vacuum. It is not possible for a particle to exceed the speed of light in a vacuum.

CLADDING – (see also canning) – the long metal tube into which fuel pellets are sealed to form elements. It is the primary barrier against the escape of radioactive fission products from the fuel.

COMMISSIONING – all the detailed testing activities involved in putting a nuclear plant into service. Decommissioning is all the operations involved in reducing the plant to a safe state after its useful working life.

CONSORTIUM – group of companies set up to undertake large nuclear plant construction projects.

CONTAINMENT – the steel and concrete building surrounding a reactor and its associated plant and equipment. It provides a further barrier against any escape of radioactivity.

CONTROL RODS – rods or plates of neutron-absorbing material, which are inserted into the core of a reactor to control or stop the fission chain reaction.

COOLANT – the liquid or gas used to transfer the heat of nuclear fission to a heat-exchanger in which steam is raised for the electrical turbo-generator. The cool liquid or gas is then returned to the reactor, round the coolant circuit.

Glossary

CORE – central region of a nuclear reactor, containing the fuel assemblies and moderator, in which the fission chain reaction takes place.

CRITICAL MASS – the minimum amount of fuel needed in the core of a nuclear reactor in order to start a self-sustaining chain reaction. When a reactor stars up it is said to 'go critical'.

CURIE (Ci) – basic unit of radioactivity. The disintegration of 3.7 3× 10^{10} nuclei per second. It was based originally on the radioactivity of 1 gramme of radium. Now being superseded by the Becquerel (Bq), equal to one disintegration per second.

DBA – (Design Basis Accident) hypothetical accident scenario used in the design of safety systems.

DECAY – radioactive emission takes place at the instant that an atom disintegrates. As the number of radioactive atoms is reduced by this disintegration, the rate of radioactive emission is said to decay.

DEPLETED URANIUM – the residual material from the enrichment process in which the proportion of the fissile uranium-235 isotope has been reduced from its natural level of 0.7 per cent to 0.2–0.3 per cent.

DEUTERIUM – a naturally occurring isotope of hydrogen. It has a proton and a neutron in the atomic nucleus and therefore a mass twice that of the most abundant isotope of hydrogen. Sometimes referred to as heavy hydrogen, a constituent of heavy water.

DISINTEGRATION – used to describe any process in which the arrangement of particles in the nucleus of an atom is disrupted, producing a change in the nature of the atom. It may happen spontaneously, in the case of a radioactive atom, or as a result of bombardment by other particles.

E.C.C.S. – (Emergency Core Cooling Systems) – back-up systems to provide injection of coolant into a reactor if there is a loss of coolant from the primary circuit.

E.D.F. – (Electricité de France) – state-owned utility responsible for electricity production in France.

EFFICIENCY – in the context of energy use, it is a measure of the amount of useful energy produced as a percentage of the primary resources consumed. In the case of a power station it is the ratio of electrical energy sent out (MWe) to thermal power within the reactor core (MWth).

Glossary

ELECTRICAL CHARGE – an electron carries a negative electrical charge designated as one unit and a positron carries a positive charge of one. A flow of electrons jumping from atom to atom in the metal of a wire is an electrical current. Atoms stripped of electrons are left with an overall positive charge associated with the positrons in the nucleus and are said to be ionized.

ELECTROMAGNETIC RADIATION – all kinds of light rays, heat rays, microwaves and radiowaves, X-rays and gamma rays. Each ray has a different characteristic wavelength.

ELECTRON – one of the basic particles of all atoms. It is very small and light and has a negative electrical charge.

ELEMENTS – basic chemical classification of atoms according to the number of positively charged protons in their nuclei (see also *fuel element*).

ENGINEERED PROTECTION – safety systems designed to provide extra layers of protection to a reactor, in addition to the intrinsic safety features, which will tend to shut the reactor down if something goes wrong.

ENRICHMENT – the physical process of increasing the concentration of the uranium-235 isotope relative to the predominant uranium-238 isotope in natural uranium. For power reactor fuel the proportion of uranium-235 is typically increased from 0.7 per cent to 3–4 per cent.

FAIL-SAFE – philosophy used in the design of safety systems. For example, control rods held up by magnets so that they will fall into a reactor core if there is a power failure.

FAST REACTOR – a reactor in which the chain reaction is sustained with the neutrons emitted at high speed – fast neutrons – from fissioning uranium or plutonium atoms. In other reactors the neutrons are slowed down by a moderating material.

FAULT TREES – detailed analysis of chains of failures that might contribute to an accident.

F.B.R. – Fast Breeder Reactor – a fast reactor with a blanket of depleted uranium around the core in which it is possible to breed new plutonium fuel at a faster rate than it is being consumed in the reactor core.

FISSION – the splitting of large atoms of uranium and plutonium, a process which releases energy.

Glossary

FOSSIL FUEL – fuels formed by prehistoric fossilization of organic materials, notably coal, oil and natural gas.

FULL SCOPE SAFEGUARDS – submission of all nuclear facilities in a country to international safeguards inspection (see also *safeguards*).

FRAMATOME – leading French company involved in design and construction of commercial nuclear power plants.

FUEL ASSEMBLY – number of fuel elements (or pins) mounted in precise grid supports for loading into a reactor.

FUEL CYCLE (NUCLEAR) – all the stages involving the fuel: uranium mining, the fuelling of a reactor, recycling of unused fuel, and radioactive waste management.

FUEL ELEMENT – nuclear fuel sealed into a tube (can) which radiates heat in the core of a reactor rather like the element of an electric kettle. In the case of the very long thin elements, sometimes referred to as fuel pins.

FUSION – the process in which two small atoms fuse together with the release of even greater amounts of energy than is released when heavy atoms are split.

GAMMA RAYS – highly penetrating electromagnetic radiation released when the nucleus of a radioactive atom disintegrates.

GENERAL ELECTRIC – the American electrical engineering company involved in the design and construction of nuclear power plants using boiling water reactors. No connection with the General Electric Company in the U.K. which is also involved in the nuclear industry as a major shareholder in the National Nuclear Corporation.

GRAPHITE – a dense black crystalline form of pure carbon which is an efficient moderator of neutrons; i.e. slows them down.

HALF-LIFE – the time required for the radioactive emission from a particular substance to 'decay' or weaken to half its original value; i.e. when half of its atoms have disintegrated. Each radioactive isotope has a characteristic half-life.

HEAT EXCHANGER – a unit in which the heat from one fluid is transferred to another. It usually takes the form of a bank of tubes carrying one (the primary) fluid in a vessel through which a secondary fluid is circulated. A steam generator is a heat exchanger in which the secondary fluid, water, boils to produce steam to drive a turbine.

Glossary

HEAVY WATER – like graphite, an efficient moderator of neutrons. It is water rich in the heavy isotope of hydrogen, deuterium. Heavy water is present in naturally occurring water at a concentration of about one part in 5000. Chemically it is identical to ordinary (light) water.

HEXAFLUORIDE (URANIUM – 'HEX') – a chemical compound of uranium and fluorine. It is a gas at low temperatures and therefore a convenient form for enrichment processes in which the uranium-235 isotope is separated.

IONIZATION – stripping of one or more electrons from the outer orbits of an atom so that the atom is left with a positive electrical charge.

INTRINSIC SAFETY – safety features imparted by the physical characteristics of a reactor. They require no human or mechanical intervention to make them effective.

IRRADIATION – the bombarding of atoms with nuclear particles to change the structure of the nucleus and produce radioactive atoms. Fuel which has been in a reactor is often called 'irradiated' because it has been bombarded with neutrons and has become radioactive.

ISOTOPES – chemically identical atoms of the same element with slightly different weights due to different numbers of neutrons in the nuclei. Isotopes are known by the total number of neutrons and protons in the nucleus – e.g. uranium-235 and uranium-238.

KILOWATT-HOUR (kWh) – dissipation of 1000 Watts of power for one hour. One unit on a domestic electricity meter represents 1 kWh.

KRAFTWERK UNION – main German company involved in the design and construction of commercial nuclear power plants.

LIGHT WATER – term used to distinguish ordinary water from heavy water.

LIQUID METAL – in effect a molten metal. Used to describe a metal such as sodium which has a relatively low melting temperature and can therefore be used conveniently as an efficient remover of the large amount of heat generated in a fast reactor.

L.M.F.B.R. – Liquid Metal Fast Breeder Reactor – a fast breeder reactor with liquid metal, usually sodium, coolant.

L.W.R. – Light Water Reactor – generic name for reactors with light (ordinary) water coolant and moderator.

Glossary

MAGNESIUM – light metallic element.

MAGNOX – alloy of magnesium used for the tubing (can) of fuel elements in some gas-cooled reactors.

MEGA – prefix for one million, e.g. megawatt (MW) = 1 000 000 Watts. Commercial nuclear reactors have outputs in the range 600 to 1300 MWe.

MODERATOR – a substance in which neutrons tend to bounce from atom to atom instead of being captured by nuclei. Moderators are therefore used to slow down ('moderate') the neutrons emitted by fissioning uranium atoms, and increase their chance of being captured by another uranium atom.

NATIONAL NUCLEAR CORPORATION – N.N.C. – leading British industrial company involved in the design and construction of nuclear power plants. Combines the former interests of competing consortia and a state shareholding.

NATURAL CONVECTION – the process in which hot fluids rise and cool fluids fall, capable of maintaining a flow of reactor coolant under favourable conditions without need for circulation pumps.

NATURAL URANIUM – uranium with the istopic concentrations found in nature, i.e. 99.3 per cent uranium-238 and 0.7 per cent uranium-235.

NEUTRON – one of the particles found in the nucleus of an atom, so-called because of its neutral electric charge. Free neutrons, released by fissioning or radioactive disintegration of atoms, are very penetrating. When they do collide with the nuclei of other atoms they are likely to cause a wide variety of changes, or transmutations, in the physical characteristics of the atoms they strike.

N.N.I. – Nuclear Installations Inspectorate – licensing body for nuclear plants in Britain.

N.R.C. – Nuclear Regulatory Commission – licensing body for nuclear plants in the United States.

PLUTONIUM – a metallic by-product in a nuclear fission reactor. It is formed by the capture of neutrons by atoms of uranium-238. Plutonium also fissions and hence can be used as a fuel. It is particularly suitable for use in fast reactors.

PRESSURE TUBE – small pressure vessels used in heavy water reactors

Glossary

to put the coolant under pressure as it flows over fuel elements in the core. Several hundred pressure tubes passing through a 'calandria' (tank) of heavy water moderator make up the core of a typical reactor.

PRESSURE VESSEL – a large vessel of steel or pre-stressed concrete containing the whole of the reactor core. Such a vessel is used to pressurize the coolant in gas-cooled and light water reactors.

PRE-STRESSED CONCRETE – a very high strength concrete produced by embedded steel cables which are stretched – pre-stressed – before any load is applied to the concrete.

P.W.R. – Pressurized Water Reactor – the widely used reactor type with high pressure water coolant and moderator and enriched uranium fuel.

QUALITY CONTROL – inspection techniques used to check the quality of manufacture of components. Quality assurance is the management of all the quality control checks to ensure that the requisite overall quality is achieved for a particular product.

RADIATION – energy given off by atoms when they are moving or changing state. Can take the form of electromagnetic waves, such as heat, light, X-rays, or gamma rays, or streams of particles such as alpha particles, beta particles, neutrons or protons.

RADIOACTIVITY – the spontaneous disintegration of some atomic nuclei with the emission of energy in the form of radiation.

RAD AND REM – units used to quantify the absorption of radiation energy in a given weight of matter (rad) or in human tissue (rem). Rems, or more often millirems (1/1000rem), are therefore used to specify radiation dose limits (see also *Sievert*).

REMOTE HANDLING – techniques used for the handling of radioactive materials behind the protection of walls which will absorb the radiation.

REPROCESSING – chemical treatment of spent fuel from a nuclear reactor to separate unused uranium and plutonium from radioactive fission product wastes. This allows recycling of valuable fuel material and reduces the volume of waste materials.

SAFEGUARDS – used in the context of fissionable materials, such as highly enriched uranium or plutonium, to describe the international system of inspection and accounting operated by the International Atomic Energy Agency to ensure that the materials are applied only for civil purposes.

Glossary

SHUT-DOWN – stopping of the fission chain reaction in a reactor by inserting neutron-absorbing control rods into the core. Also referred to as a reactor 'scram'.

SIEVERT – new unit of radiation dose that is being adopted in preference to the Rem. 1 Sv = 100 Rem.

SPENT FUEL – fuel elements taken out of a nuclear reactor after a period – usually one to three years – of useful energy production. Also referred to as 'irradiated fuel.'

STEAM GENERATOR – heat exchanger in which the secondary fluid, water, boils to produce steam to drive a turbine. Also referred to as a 'boiler'.

THERMAL REACTOR – nuclear fission reactor which uses moderated (slow or 'thermal') neutrons; e.g. Magnox, light water or Candu reactors.

THERMONUCLEAR REACTION – an alternative name for nuclear fusion reaction.

THORIUM – metallic element which can be converted by neutron irradiation into a fissionable isotope of uranium and thus to a potential new fuel.

TRANSMUTATION – the changing of atoms of one element into another element by bombardment with nuclear particles; e.g. uranium-235 into plutonium or thorium into uranium-233 by neutron bombardment.

TRITIUM – an extra-heavy isotope of hydrogen containing two neutrons and one proton in the nucleus.

U.K.A.E.A. – United Kingdom Atomic Energy Authority – nuclear research and development organization in Britain.

URANIUM – the heaviest naturally occurring element.

VITRIFICATION – incorporation of waste materials into a solid ingot of glass.

WASTE MANAGEMENT – all the procedures for treatment of radioactive waste materials, safe storage and eventual disposal.

WATT (W) – unit of power. A typical light bulb consumes 100 Watts. A one-bar electric fire consumes 1 kilo-Watt (1000 Watts). Nuclear reactors produce 600 to 1300 mega-Watts (million Watts) of electricity.

Glossary

WESTINGHOUSE – the American electrical engineering company involved in the design and manufacture of nuclear power plants using pressurized water reactors.

YELLOW CAKE – refined chemical compound of uranium, the form in which uranium is usually shipped from the mine to the nuclear fuel manufacturer.

ZIRCONIUM – metallic element. An alloy of zirconium known as Zircaloy is extensively used for the canning, or cladding, of nuclear fuel elements.

Index

A.B. Atomenergi, Sweden, 183
Absorber, 24
Accidents do happen, 9, 87–89
Accidents, industrial, 8
Actinides, 124
Advanced converters, 128
Advanced reactors, 128–151
Advanced gas-cooled reactor (A.G.R.), 34, 37–40, 156
Aerodynamic enrichment, 56
Alpha radiation, 69
Alpha wastes, 106
Argentina, 193
As low as readily achievable, 72
Atomic pile, 22
Atomic Energy of Canada Limited (A.E.C.L.), 177
Atomic Energy Commission, U.S., 165
Atoms for Peace, 16, 91, 152
Austria, 101, 189
A.E.G., Germany, 173
A.S.E.A.–Atom, Sweden, 183

Babcock & Wilcox, U.S., 166
Backfilling, 121
Background radiation, 71
Barnwell, U.S., 172
Barriers, 119
Basque Separatists, 103
Becquerel, 70
Belgium, 186
Beloyarsk, U.S.S.R., 130
Bentonite, 121
Berkeley, U.K., 155
Beta radiation, 69
Beta-gamma wastes, 106
Biblis, Germany, 174
Binding energy, 21, 148
Blanket, 46, 133

Boilers, 35
Boiling water reactor (B.W.R.), 31–34
Borosilicate glass, 117
Bradwell, U.K., 155
Brazil, 194
Breeding, 6, 47, 98, 133
Britain, 153–158
British Nuclear Fuels Limited (B.N.F.L.), 157
Brown Boveri
  Germany, 174
  Switzerland, 188
Bruce, Canada, 178
Burn-up, 30
But what if? 9

Cascade, enrichment, 54
Cadarache, France, 161
Calciner, 117
Calder Hall, U.K., 153
California, 102, 169
Can nuclear reactors explode? 24
Canada, 177–180
Cancer risk, 71, 73
Candu reactor, 40–43, 147, 177
Cap La Hague, France, 163
Capenhurst, U.K., 157
Carbon dioxide, 26
Carter, U.S.A., 94
Central Electricity Generating Board (C.E.G.B.), 154
Centrifuge enrichment, 55, 157
Cesium-137, 124
Chain reaction, 21
Chemical enrichment, 56
Choice of reactor, 155, 160, 166, 173, 177
Cladding, 26
Cladding tube, 60

Index

Classification of waste, 106–107
Clean-up systems, 79
Coated particle fuel, 138
Cogema, France, 162
Cogeneration, 141
Combustion Engineering, U.S.A., 166
Comecon countries, 196–198
Commercial Demonstration Fast Reactor (C.D.F.R.), 130
Commissariat à l'Energie Atomique, France, 159
Committee for Assurances of Supply, 97
Construction permit, 81
Containment, 77
Control and instrumentation, 77
Control rod, 24
Coolant, 26
Cooling ponds, 62
Core catchers, 136
Cosmic radiation, 72
Costs, 18, 123
Creusot Loire, France, 160
Creys-Malville, France, 130
Criticality, 24
Curie, 70
Czechoslovakia, 198

Decay heat, 75
Decay, radioactive, 14
Decommissioning of power stations, 126–127
Defence in depth, 8
Delayed neutrons, 25
Depleted uranium, 46, 54
Design basis accident, 82
Deuterium, 26, 149
Developing world, 190–196
Dissolving fuel, 65
District heating, 142
Divergence, 24
Do we have enough uranium? 6
Do we need nuclear energy? 4
Dose, radiation, 71
Dual and special purpose reactors, 141–143
Dungeness B, U.K., 156

Economics, 18–19
   Canada, 178
   fast reactors, 136

France, 164
fusion, 150
Germany, 174
nuclear ships, 145
small reactors, 147
U.K., 158
U.S.A., 167
Economy of scale, 155
Efficiency, 2
Egypt, 195
Eisenhower, U.S.A., 91
Electricité de France (E.d.F.), 159
Electromagnetic radiation, 69
Emergency core cooling, 31, 76
E.N.E.L., Italy, 186
Energy, 4
   needs, 3–5
Enrichment of uranium, 22, 46, 53–57, 92, 171, 175, 183
Eurochemic, Belgium, 187
Eurodif, France, 162
Experimental Breeder Reactor I (E.B.R. I), 128

Fail-safe, 8, 74
Falldin, Sweden, 100
Fast breeder reactor, 47, 97
Fast neutrons, 22
Fast reactor, 6, 23, 27, 46, 129–134
   economics, 136
   fuel, 66, 133
   safety, 167
Fault trees, 83
Fear, 68
Fermi, Enrico, 22
Film badge, 74
Final disposal, 119
Finland, 188
Fission, 21
Fission chain reaction, 21–24
Fission products, 14, 46
   decay, 124
   wastes, 64
Flasks, transport, 62
Flux, neutron, 129
Forecasts, 3
Framatome, France, 160
France, 159–165
Fuel, 25, 44
   assembly, 28, 59
   cycle, 44, 95

Index

element, 26, 71
fabrication, 59–61
flasks, 62
pellets, 60
pins, 133
re-fabrication, 66–67
Fuelling machine, 37
Full scope safeguards, 94
Fusion, 147–151

Gamma radiation, 69
Gas turbine, 140
Gas-cooled reactors, 27, 34–40
Gaseous diffusion, enrichment, 53, 162
Gaseous waste, 106, 110
Gel precipitation of mixed oxides, 66
General Electric, U.S.A., 166
Geological disposal, 113, 119
German local politics, 126
German Reprocessing Company, 176
Germany, Federal Republic, 172–176
Giscard d'Estaing, France, 160
Glassification, 117
Global energy requirements, 5
Gorki, U.S.S.R., 143
Granite, 121
Graphite, 22, 27, 138
Gray, 70

Half-life, 14
Head end, reprocessing, 64
Health physics, 73
Heat generating wastes, 123
Heat only reactors, 142
Heat transfer, 26
Heavy water, 26, 42, 57–59
production, 57–59, 179
reactors, 27, 40–43
Helium, 26
High level waste, 106, 114–116
High temperature reactors, 137–141
High temperature gas-cooled reactor (H.T.G.R.), 34
Hitachi, Japan, 181
How much energy do we need? 2
How safe is safe enough? 82–89

Ice breakers, 144
Incident reporting, 89
India, 192
Indian explosion, 94

Industrial organization, 100
Institute of Nuclear Power Operations, 170
Interim storage, 62, 115
Intermediate level waste, 106, 112
International Atomic Energy Agency (I.A.E.A.), 16, 49, 92, 147
International Commission on Radiological Protection (I.C.R.P.), 7, 72, 111
International Nuclear Fuel Cycle Evaluation (I.N.F.C.E.), 44, 94, 114
Ion exchange, 51
Iraq research reactor attack, 96
Irradiated fuel, 61
Is nuclear energy economical? 18
Is nuclear energy safe? 7
Isotopic separation, 53–57
Israel attack on Iraqi reactor, 96
Italy, 102, 185

Japan, 180–183
Jet nozzle enrichment, 56, 176

Kraftwerk Union, Germany, 174
Kreisky, Austria, 101

Laser enrichment, 56
Leach rates, 118
Lenin, nuclear icebreaker, 144
Licensing, 80
Light water, 27
reactors (L.W.R.), 27–34
Linear dose-effect relationship, 71
Liquid waste, 106
Lithium, 149
Local politics, 101–103
London Suppliers' Club, 94
Long-term prospects for fusion reactors, 147–151
Loop-type fast reactor, 133
Loops, 28
Low-level waste, 106, 111
Lucens, Switzerland, 9

Magnox, 26
reactor, 35–37, 146
Manhattan project, 54
Marcoule, France, 117, 163
Mexico, 195
Mining, 51
wastes, 107

212

Index

Mitsubishi, Japan, 181
Mitterrand, France, 160
Mixed oxide fuel, 66, 133, 186
Mixer settlers, 63
Moderator, 22, 26
More from less, 2
Mutsu, nuclear ship, 144

National politics, 99–101
National Nuclear Corporation, U.K., 155
Natural radiation, 72
Natural reactor, 23
Natural uranium, 23, 40
Need for energy, 2
Netherlands, 189
Neutron, 22
 activation, 70, 150
 radiation, 69
Non proliferation, 91–96
Non Proliferation Treaty (N.P.T.), 16, 93
Nuclear
 explosion, 24
 fuel, 44–67
 industry worldwide, 152–198
 reactors, 21–43
 safety, 68–90
 safety measures, 74–80
 ships, 144
Nuclear Energy Agency (N.E.A.), 49, 187
Nuclear Regulatory Commission, U.S., 170
N.P.T. review conference, 93

Ocean
 disposal, 123
 dumping, 113
Odessa, U.S.S.R., 142
Oil crisis, 4
On-load fuelling, 37, 42
Once-through fuel cycle, 47
Ontario Hydro, Canada, 177
Operating licences, 81
Operational safety, 89–90
Organic coolants, 26
Otto Hahn, nuclear ship, 144
Oxide reprocessing, 64

Pebble bed reactor, 138
Pechiney Ugine Kuhlmann, France, 162

Pickering, Canada, 178
Plasma, fusion, 148
Plutonium, 6, 13, 46, 63, 114, 129, 133
 politics, 96–101
 storage, 97
Political extremists, 103–104
Politics of nuclear energy, 91–104
Pool-type fast reactor, 131
Population, 4
Post-industrial society, 1
Power Reactor & Nuclear Fuel Development Corporation, Japan, 183
Pre-stressed concrete pressure vessel, 35, 39
Pressure tubes, 40
Pressure vessel, 28
Pressurization, 28
Pressurized water reactor (P.W.R.), 28–34
Primary circuit, 28
Probability analysis, 83
Process heat, 6, 140
Proliferation, 16–18
Prospect for small land-based plants, 146–147
Prospecting, uranium, 51
Pulsed columns, 63

Quality control, 81
Questions, 1–20

Rad, 70
Radiation, 68–72
 danger of, 12
 dose, 71
 units, 70
Radiological protection, 72–74
Radon, 52
Reactor, 21
 types, 25–43
Reagan, U.S.A., 170
Recycle, 7
Redundancy, 75
Regulation, 80–82
Rem, 70
Reprocessing, 44, 63–66, 92, 95, 114, 157, 163, 172, 176, 183, 187
Residual heat, 61, 75
Risk analysis, 84
Risks, 10
Rock salt, 121

Index

Safeguards, 16, 92
Safety, 7-9, 68-90
  reports, 80
Salt mine, 113
Savannah, nuclear ship, 144
Scram, 24
Seismic design, 75
Sellafield, U.K., 157
Separative work units, 57
Shallow land burial, 113
Ship propulsion reactors, 143-146
Shut down, 24
Siemens, Germany, 173
Sievert, 70
Sizewell B, U.K., 156
Small reactors, 146-147
Sodium, 26, 130, 135
Solid waste, 106
Solidification of high-level wastes, 116-118
Solvent extraction, 51, 63
South of Scotland Electricity Board, 154
South Korea, 193
Soviet Union, 130, 144, 196
Spain, 103, 187
Spent fuel, 46, 61, 114
  storage and transport, 61-63
Springfields, U.K., 158
Stainless steel cladding, 38
Steam generator, 28
Storage
  of fuel, 62
  of high-level liquid wastes, 116
  of spent fuel, 115
Sweden, 100, 183
Switzerland, 188

Tailings, mine, 52, 107
Taiwan, 193
Temperature coefficient, 25, 75, 135
Terrorism, 17, 103
Thermal efficiency, 38, 137
Thermal neutrons, 22
Thermal reactors, 6, 27, 46
Thermal recycle, 47, 66
Thorium cycle, 47, 129, 138
Three Mile Island, U.S.A., 9, 87, 169
Timescale and cost for waste disposal, 123-126
Tricastin, France, 163
Tokamak, 148

Toshiba, Japan, 181
Tritium, 149

U.K. Atomic Energy Authority, 154
U.S. Non Proliferation Act, 94
United States, 165-172
Uranium, 21, 179
  carbide, 138
  conversion, 53
  enrichment, 53-57
  hexafluoride, 53
  mine tailings, 134
  mining, 51-53
  oxide, 60
  requirements, 6
  reserves, 7
  resources, 48-53
  supplies, 7-9
Uranium-233, 47, 133, 138
Uranium-235, 21, 46
Uranium-238, 22, 46
Urenco, 158, 175

Vitrification, 117
Vorenezh, U.S.S.R., 143
Vortex tube enrichment, 56

Walk-away safety features, 135
Waste, 10-16, 46
  arising at power stations, 110-114
  disposal, 119
  from fuel production, 107-110
  heat, 141
  management, 67, 82-83, 105-127
  production, 107, 108, 109
  repositories, 148
  solidification, 117
Western Europe, 183-190
Westinghouse, U.S.A., 166, 171
What about proliferation? 16
What about waste? 10
What sort of world? 1
Who do you believe? 19
Windscale, 9
  accident, 87

X-rays, 69

Yellow cake, 51

Zircaloy, 26